An Introduction to Verification of Visualization Techniques

Synthesis Lectures on Visual Computing
Computer Graphics, Animation, Computational
Photography, and Imaging

Editor
Brian A. Barsky, *University of California, Berkeley*

This series presents lectures on research and development in visual computing for an audience of professional developers, researchers and advanced students. Topics of interest include computational photography, animation, visualization, special effects, game design, image techniques, computational geometry, modeling, rendering, and others of interest to the visual computing system developer or researcher.

An Introduction to Verification of Visualization Techniques

Tiago Etiene, Robert M. Kirby, and Cláudio T. Silva

ISBN: 9781627058339   paperback
ISBN: 9781627058346   ebook

DOI 10.2200/S00679ED1V01Y201511CGR022

A Publication in the Morgan & Claypool Publishers series
SYNTHESIS LECTURES ON VISUAL COMPUTING: COMPUTER GRAPHICS, ANIMATION, COMPUTATIONAL PHOTOGRAPHY, AND IMAGING

Lecture #22
Series Editor: Brian A. Barsky, *University of California, Berkeley*
Series ISSN
Print 2469-4215   Electronic 2469-4223

An Introduction to Verification of Visualization Techniques

Tiago Etiene
Modelo Inc.

Robert M. Kirby
University of Utah

Cláudio T. Silva
New York University

SYNTHESIS LECTURES ON VISUAL COMPUTING: COMPUTER GRAPHICS, ANIMATION, COMPUTATIONAL PHOTOGRAPHY, AND IMAGING #22

ABSTRACT

As we increase our reliance on computer-generated information, often using it as part of our decision-making process, we must devise tools to assess the correctness of that information. Consider, for example, software embedded on vehicles, used for simulating aircraft performance, or used in medical imaging. In those cases, software correctness is of paramount importance as there's little room for error. Software verification is one of the tools available to attain such goals. Verification is a well known and widely studied subfield of computer science and computational science and the goal is to help us increase confidence in the software implementation by verifying that the software does what it is supposed to do.

The goal of this book is to introduce the reader to software verification in the context of visualization. In the same way we became more dependent on commercial software, we have also increased our reliance on visualization software. The reason is simple: visualization is the lens through which users can understand complex data, and as such it must be verified. The explosion in our ability to amass data requires tools not only to store and analyze data, but also to visualize it.

This book is comprised of six chapters. After an introduction to the goals of the book, we present a brief description of both worlds of visualization (Chapter 2) and verification (Chapter 3). We then proceed to illustrate the main steps of the verification pipeline for visualization algorithms. We focus on two classic volume visualization techniques, namely, Isosurface Extraction (Chapter 4) and Direct Volume Rendering (Chapter 5). We explain how to verify implementations of those techniques and report the latest results in the field of verification of visualization techniques. The last chapter concludes the book and highlights new research topics for the future.

KEYWORDS

visualization, verification, isosurfaces, volume rendering, geometry processing, verifiable visualization

Contents

x

Preface

The term *verification* has become ubiquitous in both the computer science and engineering communities as denoting a process that somehow convinces the user that verified tools, whether those be circuits, algorithms, implementations, etc. are more safe, accurate, or complete than other tools that have not been verified. Although the term verification has a common root usage within both communities, it has evolved to mean something specific to each subarea of computer science and of engineering. For instance, within computer science, the verification of a circuit denotes either the exhaustive testing or proof that under all possible inputs, the circuit will produce the correct (specified) outputs. Similarly, for software, verification relates how well an implementation represents the behavior of its specification under all possible inputs. Within the engineering world, verification takes on a different, more nuanced meaning. One assumes that there exists an "exact solution" or "exact representation" resulting from the solution of a mathematical system of equations. In all but the most trivial circumstances, this exact solution is not attainable, and approximate solutions must be formed. The process of quantifying how well a numerical scheme or representation approximates the exact solution is referred to as verification. Verification may involve looking at how well (or quickly) an approximate solution converges (in an appropriate norm) to the exact solution, or may involve identifying features or invariants of the solution that should be maintained regardless of the approximate representation. As visualization models, algorithms and implementations lie at the interface of these two branches, what does it mean to produce *verifiable visualizations*?

This question motivated the research work that has become the foundation of this book. To answer such a broad question, we started as most researchers would: by examining a concrete example in which our ideas could be refined. We started with isosurface extraction. Many tests and a few software bugs (which our process found) later, we realized that not only were our results worth communicating to the community, but that there was much work still to do. We moved to verifying different techniques used within the visualization community—in turn learning new things along the way. We began to appreciate that verification is a process, and that articulating the guiding principles of that process was itself a contribution to our community. The various papers we reference outline the specific contributions of our work. This book is meant to make that work accessible to the general reader in a pedagogical way. We hope the reader will take away not just a particular technique, but a way of approaching and testing visualization algorithms and their implementations. In the end, we hope that all successful visualization techniques will produce verifiable visualizations.

Any work of this size and scope has benefitted by many people both indirectly and directly. We wish to thank our collaborators that helped to shape this work, in particular Luis

Gustavo Nonato, Carlos Eduardo Scheidegger, Julien Tierny, Thomas J. Peters, Valerio Pascucci, Daniel Jönsson, Timo Ropinski, João Luiz Dihl Comba, Anders Ynnerman, Lis Custódio, and Sinésio Pesco. We also thank the various faculty and students at the SCI Institute (University of Utah) with whom we sharpened our ideas. In addition, we would like to thank the various Federal Funding Agencies that have supported our research efforts over the years. The papers we reference which are co-authored by us detail those acknowledgements. Lastly, we would like to thank our spouses, without whose patience and encouragement we would probably not have made it this far.

Tiago Etiene, Robert M. Kirby, and Cláudio T. Silva
December 2015

CHAPTER 1

Introduction

The scientific method, as introduced by Aristotle, was formulated around the idea of postulating a model (or ideal in the Platonic sense) of natural phenomenon, making observations to validate one's model, and correcting the model based upon discrepancies between the phenomena and nature. Sir Francis Bacon is attributed with extending this process to include the idea of the controlled experiment. No longer were the scientists limited to passively observing the world around them to deduce the correctness of the model. This gave rise to the idea of devising controlled experiments designed to evaluate the correctness of the hypothesis in a systematic manner. This systematic process allowed the model to evolve based upon the lessons learned through the experiment. The late Microsoft researcher Dr. Jim Gray argued in [51] that since Bacon, there have been four paradigms of scientific discovery: experimental science, theoretical science, computational science, and data science. The first two of these paradigms reigned from the time of Bacon through the early part of the 20th century. Since the advent of computing, the latter two paradigms have risen to prominence.

With the advent of modern computing, the first of the two paradigms, called simulation science, has emerged. In this paradigm, the *experiment* now employed within the scientific method consists of the computational solution of the model. The scientific method underlying simulation science is composed of the following stages.

- **Scientific Problem of Interest (*Problem Identification*).** Statement of the scientific or engineering problem of interest. Questions should be developed in such a way that quantifiable metrics for determining the level of success of the simulation science endeavor can be evaluated.

- **Modeling.** The development of a model that abstracts the salient features of the problem of interest in such a way that exploration and evaluation of the model allows an answer to the questions specified concerning the problem of interest. Modeling techniques include, but are not limited to, deterministic or probabilistic, discrete or continuous mathematical models. Means of validating the model (determining the error introduced due to the model abstraction of the real phenomenon) should be established.

- **Computation.** The generation of algorithms and implementations that accurately and efficiently evaluate the model over the range of data needed to answer the questions of interest. This simulation of the physical phenomenon by computational expression of the model provides the experiment upon which the simulation scientific method hinges.

- **Evaluation.** The distillation and evaluation of the data produced through computational simulation to answer the questions of interest and to provide quantifiable determination of the success of the experiment. Methods such as scientific visualization provide a means of tying the simulation results to the problem of interest.

The use of simulation science as a means of scientific inquiry is increasing at a tremendous rate. It is now used in a diversity of fields such as aircraft and automobile design, climate modeling, and drug design. The process of mathematically modeling physical phenomena, experimentally estimating important key modeling parameters, numerically approximating the solution of the mathematical model, and computationally solving the resulting algorithm has inundated the scientific and engineering worlds, allowing for rapid advances in our modern understanding and utilization of the world around us. But at the end of the day, how do we know that our computational results are "right?" That is to say, how do you know that they should be trusted, or to what level should they be trusted?

This book was motivated by the use of visualization as a means of evaluation in the third paradigm, but it is relevant to both the simulation science paradigm and the data science paradigm. Visualization is often employed as part of the simulation science pipeline. It is the lens through which scientists often examine their data for deriving new science, and the lens used to view modeling and discretization interactions within their simulations. In [24], we proposed that visualization itself must be explicitly considered with similar scrutiny as other parts of the pipeline if it is to be used as part of this scientific process.

Examining the pipeline itself and understanding what procedures need to be in place to help ensure that its results are meaningful and reliable falls under the general heading of "V&V" – Validation and Verification—within the simulation science community. We define validation and verification as follows (definitions are taken directly from [2]).

- **Validation.** The process of determining if a mathematical model of a physical event represents the actual physical event with sufficient accuracy.

- **Verification.** The process of determining if a computational model obtained by discretizing a mathematical model of a physical event and the code implementing the computational model can be used to represent the mathematical model of the event with sufficient accuracy.

Based upon these two definitions, it is easy to see that fundamental to simulation science is the idea of the "error budget"—those assumptions and approximations that introduce error (or approximations) into the simulation process and their corresponding impact (or cost) on the scientific pipeline. Quantification, and ideally elimination, of modeling errors (those errors introduced through the choice of a mathematical model to describe observable data), approximation error (those errors introduced in the numerical computation of solutions of the model), and uncertainty errors (those errors due to variation in model parameters) are critical components of the scientific process. They allow scientists to judiciously evaluate which component of the process described above (e.g., modeling, numerical approximations) requires refinement in comparison with the real phenomenon of interest. Over the last 40 years, tremendous effort has been exerted in the pursuit of numerical methods that are both *flexible* and *accurate*, hence providing sufficient fidelity to be employed in the numerical solution of a large number of models and sufficient quantification of accuracy to allow researchers to focus their attention on model refinement and uncertainty quantification. It is in light of this that the verification process currently used in simulation science has been solidified; it is a means of "proving the mettle" of the computational and mathematical model [2].

The verification process is commonly partitioned into two areas recognizable to most visualization researchers: solution verification and code verification. In solution verification, effort is directed toward assuring the accuracy of the input data, estimating the numerical approximation error due to discretization, and assuring the accuracy of the resulting simulation output data. In code verification, effort is directed toward finding and removing source-code mistakes and finding and removing (numerical) algorithmic errors. When these two forms of "debugging" are accomplished, they allow researchers not only to correct and refine their scientific tools, but also to build a confidence in the design and handling of the scientific tool and the corresponding results it produces.

When these results are then to be used in the scientific setting, differences between computational and experimental results can be examined in light of the assumptions that were employed in the model generation and simulation. If visualization is the lens through which simulation scientists view their data, is that lens free of flaws? Is it possible that visual discrepancies between simulation and experimental results could be due to assumptions and approximations built into the visualization method? Are the visualization techniques designed based upon (and, in particular, to respect) properties of the model and the simulation used to generate the data being visualized? To place visualization firmly within the scientific process, it must undergo the same level of rigorous analysis.

VERIFIABLE VISUALIZATION

Data visualization has become an indispensable means of presenting data due to its ability to succinctly summarize and support ideas and concepts that are being examined or presented. A basic premise of visualization is that visual information can be processed at a much higher rate than raw numbers and text. As the cliché goes, "A picture is worth a thousand words" [23]. Visualization techniques and systems [22, 43, 44, 48, 55] have thus emerged as a key enabling technology in this endeavor: helping people explore and explain data by allowing the creation of both static and interactive visual representations.

Visualizations libraries such as Kitware's VTK contain a very large number of highly complex visualization algorithms with thousand of lines of code implementing them. The most powerful of these algorithms are often based on complex mathematical concepts, e.g., Morse-Smale complex [9], spectral analysis [45], and partial differential equations (PDEs) [3]. Robust implementations of these techniques require the use of nontrivial techniques (e.g., simulation of simplicity [10], linear systems solvers [6], and Delaunay meshing [50]). The overall complexity and size of these datasets leave no room for inefficient code, thus making their implementation even more complex. On top of all this, hardware keeps changing quickly and many platforms need to be supported. In particular, the use of GPUs just adds to the overall complexity. *Given all this complexity, an important question that must be asked is whether the derived visualizations are correct—both mathematically and perceptually.*

As we become more reliant on computational algorithms and systems in our day-to-day lives, there is an increasing need to develop metrics by which we can attest to the "quality" of the hardware and software components that we employ. Good design specifications are not enough as many stages of development exist between the conceptual design phase and the finished product. Furthermore, system complexity has been increasing rapidly, making it easy for "bugs" to creep inside even the most carefully designed and implemented codes.

The issue of guaranteeing correctness of complex systems has been studied in different contexts and it continues to be an active area of research [17, 18, 26]. In computer science, such considerations have proved to be important in areas such as circuit and software design. In the context of engineering, such considerations are important in the modeling and simulation of physical phenomena. Although the specific processes used in these two areas can vary significantly, they have at their core a common root paradigm, that of *validation* and *verification.*

Despite the fact that visualizations are widely used, the problem of verifying visualization algorithms and techniques has been largely overlooked [16, 20, 24, 54, 60]. Although there are ad hoc solutions for testing implementations, no technique provides a commonly accepted framework for verifying the (mathematical) accuracy, reliability, and robustness of visualization tools.

As mentioned earlier, this is distinct from, but intimately related to, questions of perception and visual representation efficacy or correctness. In fact, there has been substantial anecdotal evidence of visualization techniques whose flaws caused the misinterpretation of the underlying phenomena. Some researchers have even argued that the problem is so acute that users should avoid third-party visualization tools due to their concern about potentially incorrect results [24].

But what does it mean to produce *verifiable* visualizations? This book presents our efforts in trying to formally define a process. We start the book with a brief introduction to visualization in Chapter 2. This is followed by a brief description of validation and verification in simulation science in Chapter 3. Through a simple example, we illustrate the main steps necessary to implement a V&V pipeline. The next two chapters form the technical core of the book where we undertake a formal verification study of the correctness of isosurfacing and volume rendering techniques. In Chapter 4, we introduce the tools necessary for the verification of geometrical properties of isosurface extraction algorithms. In Chapter 5, we introduce the principles of verification of volume rendering algorithms. The final chapter, Chapter 6, provides concluding remarks and highlights new research topics for the future.

CHAPTER 2

Visualization in the Real World

In this chapter, we briefly introduce the field of "visualization" and explain the goal of this book—to introduce techniques capable of answering the question "How do I know that the visualization I see is correct?" We will focus on the importance of visualization and why it should be verified. We start by introducing the many flavors of visualization, followed by a brief history and applications. Then, we explain the typical pipeline used in scientific visualization, covering the process of data acquisition, filtering, and mapping. Next, we explain how different errors can affect that pipeline. We list some of the many error sources that hinder visualizations and present an historical account of the pursuit of the correctness of a well-known visualization *technique*. Last, we introduce some of the current practices within the visualization community and demonstrate the need for more tools for verifying visualizations.

2.1 VISUALIZING DATA

We live in the age of data, an age defined by the use of data to augment our capacity to understand and solve real-world challenges. In medicine, for instance, data helps medical diagnosis. In business, customer data is a rich source of information about customer tendencies and needs. For the individual, data can provide insights into one's health via sensors that measure weight and blood pressure. These are only a small fraction of the applications that benefit from understanding data. Nevertheless, data serves no purpose if it cannot be analyzed. As our capacity to amass and store data grows, so does the need to analyze, explore, extract meaning, and present that data to empower its users. The multitude of sources—the census, weather, medicine, satellites, numerical simulations, wearables, to name a few—adds to the problem. As the goal is to learn as much information from the data as possible, a combination of statistics, computer software, and data visualization is essential to allow us to gain insights from data.

Even though this combination of tools is taken for granted nowadays, it is not straightforward to realize that "data visualization" should be part of this tool set. In fact, there was a time when statistical graphics did not enjoy the prestige that it does today, as it was thought of as just a means "for showing the obvious to the ignorant" [52], or not as rigorous as numerical calculations. By visualizing data, the user is exposed to features that may be hard to understand otherwise. The Anscombe's quartet is a classic example advocating the need to visualize data [1]. The quartet is composed of four distinct sets of (x, y) pairs, whose mean, variance, linear regression, and other metrics are nearly identical (see Table 2.1). In other words, by these measures, the datasets are also nearly identical. By using a simple plot, however, one can clearly see that the opposite is true

Table 2.1: The Anscombe's quartet. Each dataset possesses nearly identical mean, variance, linear regression, and other metrics. By these metrics, the four datasets are also nearly identical.

Dataset I		Dataset II		Dataset III		Dataset IV	
x	y	x	y	x	y	x	y
10	8.04	10	9.14	10	7.46	8	6.58
8	6.95	8	8.14	8	6.77	8	5.76
13	7.58	13	8.74	13	12.74	8	7.71
9	8.81	9	8.77	9	7.11	8	8.84
11	8.33	11	9.26	11	7.81	8	8.47
14	9.96	14	8.1	14	8.84	8	7.04
6	7.24	6	6.13	6	6.08	8	5.25
4	4.26	4	3.1	4	5.39	19	12.5
12	10.84	12	9.13	12	8.15	8	5.56
7	4.82	7	7.26	7	6.42	8	7.91
5	5.68	5	4.74	5	5.73	8	6.89

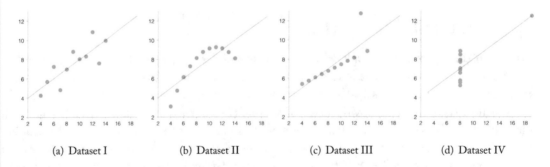

    (a) Dataset I   (b) Dataset II   (c) Dataset III   (d) Dataset IV

Figure 2.1: Scatterplot of the Anscombe's quartet shown in Table 2.1. The line in the figure is the linear regression of each dataset.

(see Figure 2.1). By visualizing the data, the user is able to make better decisions regarding, for instance, the best mathematical model for adjusting the data (Figure 2.1 is not well represented by a linear model) to detect outliers, clusters, and potentially other information not easily accessible from the data table and other statistical summaries. As the cliché goes, "a picture is worth a thousand words," or "a thousand numbers."

2.1.1   PRECURSORS OF MODERN VISUALIZATIONS

Of course, Anscombe's quartet is not the first successful display of quantitative data. Although the field of data visualization is relatively new, the use of images to depict information is much older.

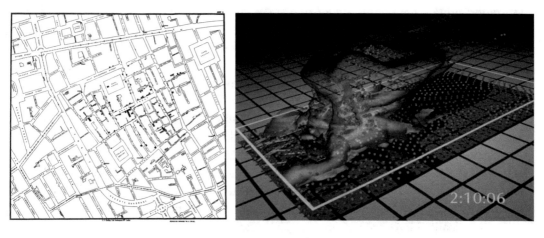

(a) Cholera Map by John Snow (1854)   (b) Severe Storm [59]

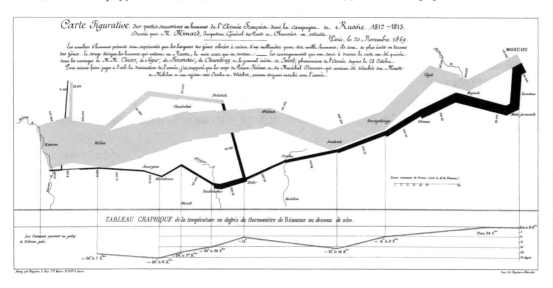

(c) Napoleon's March by Charles Minard (1869)

Figure 2.2: Classic visualization examples.

One of the first known uses of data visualization is a time-series depiction of heavenly bodies, dating back to the 10th century [14]. Since then, many visualizations have achieved the hall of fame. In particular, we mention three important cases, the first of which is John Snow's cholera map—a simple yet powerful visualization of cases of cholera in central London (see Figure 2.2(a)). A map of central London was enriched with stacked bars representing deaths from cholera and dots representing water pumps. By looking at the data in a map, John Snow was able to observe that cholera cases were concentrated around one water pump in central London, which was then removed in order to avert the crisis. The second example is the seminal Napoleon's March by Charles Minard (see Figure 2.2(c)), a depiction of a multivariate dataset describing the shrinkage of the French emperor's army as they marched toward Moscow. A third example, this one contemporary, is the "Study of a Numerically Modeled Severe Storm" (see Figure 2.2(b)) by Wilhelmson et al. [59]. The authors developed an animation depicting the intricacies of a storm. The image was generated by sensory data from storms and the numerical solution of partial differential equations. A redesigned version of the same image can be found in Tufte [53]. Visualization helps users make sense of great amounts of data and can provide a basis for decision-making. The interested reader will find more information about the aforementioned cases and, more broadly, the history of data visualization in the excellent book chapters by Friendly [14] and Tufte [52].

Visualizations have became commonplace and within reach of the general public. Newspapers, for instance, make extensive use of visualization to improve storytelling and communicate findings in a variety of subjects, from sports to international affairs. In addition, many tools have been developed specifically for using visualizations to solve problems. Tools such as NameVoyager [57]—a web-based visualization of the number of babies with a user-selected name born per year—have great social appeal. Another well-known example is ManyEyes, a collaborative platform for building visualizations. ManyEyes allows users to upload their own data or work with publicly available data. Such visualizations go beyond the typical task-oriented approach toward a more social approach [19, 57], where users engage with each other during the process of data exploration.

We have not yet made any distinction among the many subfields of visualization, notably, information visualization, visual analytics, and scientific visualization. The visualization techniques developed within each of these communities can greatly enhance the capabilities of particular users. For instance, techniques developed for the medical community can be very different from those developed with newspaper readers in mind. Each group will have specific data challenges that must be dealt with—for example, the data type, data size, data processing choices, and display requirements. The scientific computing community, for example, uses computer capabilities to understand phenomena such as fluid flow, weather prediction, and combustion simulation. By the end of the 1980s, the advances in computer power and numerical techniques allowed the scientific computing community to generate huge amounts of data from numerical computations. Nevertheless, the results were hard to evaluate—and even visualize—because there was no technology available for presenting the ever-increasing amount of information. As the Anscombe's quartet exemplifies, analyzing the data in terms of numbers alone is not enough. In response to the lack of appropriate technology and techniques, a panel of experts wrote a report on the importance of visualization for scientific computing and encouraged the National Science Foundation (NSF) to develop this new field, which was then called Visualization in Scientific Computing

(ViSC). Since then, ViSC, known today as scientific visualization (SciVis), has matured and is now part of the pipeline of users from many disciplines across science [34].

The NSF report was an important milestone in the development of the visualization community. For over two decades, conferences have been dedicated exclusively to research of new visualization techniques applied to a variety of domains. In particular, IEEE VIS (Visual Analytics, Information Visualization, and Scientific Visualization) and EuroVis conferences have been in the forefront of the visualization field, producing some of the most exciting research in the area. The first edition of the IEEE VIS (known at the time as IEEE Visualization) was held in 1990, with 54 papers published in the conference proceedings. The IEEE VIS held in 2014 included over 1000 attendees, and 134 papers were published in the conference proceedings.

Our focus is on verification of visualization, a subfield of scientific visualization (see Figure 2.3). In particular, we will describe a tool to investigate the correctness of two widely used visualization techniques available in the scientific visualization literature. We start by describing a simplified visualization pipeline.

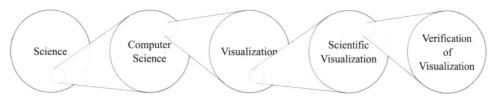

Figure 2.3: We are interested in the verification of visualization algorithms and implementations, a subfield of scientific visualization.

2.2   VISUALIZATION PIPELINE

Consider the following scenario: a man experiencing pain in his head consults a doctor. After several physical examinations, the doctor recommends the patient have his head scanned by a computerized tomography (CT) machine. The doctor hopes to find the cause of the pain by exploring a 3D model of the patient's head. What is the process behind building that model?

The typical steps involved in the visualization of scientific data are the following: *data acquisition, filtering,* and *mapping.* In our example, data acquisition happens while the patient lies down on a table; the CT machine takes several x-rays, from multiple angles, as illustrated in Figure 2.4(a). This step is perhaps the most familiar to the general public, as the patient must interact directly with the machine that generates the images. Nevertheless, by themselves, these images do not provide an idea of the internal structures of the patient's head and cannot be used for medical diagnosis. In the filtering step, the data is manipulated and prepared before generating a 3D model. The preparation involves the reconstruction of the region of space through which the x-rays travel. Each tissue layer inside the patient's head has a different x-ray attenuation coefficient. Thus, to reconstruct a 3D image, we need to build a 3D field, such that an x-ray

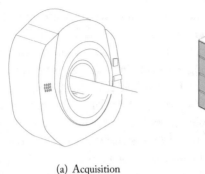
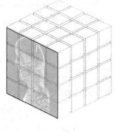

(a) Acquisition   (b) Filtering   (c) Mapping

Figure 2.4: Visualization pipeline.

going through it would be attenuated by the same amount as detected by the CT scan (see Figure 2.4(b)). Other examples of filtering include downsampling (for instance, moving data from 16-bit to 8-bit precision), noise removal, and smoothing. The last step, data mapping, can be done via volume rendering the 3D field or extracting polygonal meshes that represent the internal structures (Figure 2.4(c)). (We will review both algorithms in Chapters 4 and 5.) After that, the doctor can evaluate a 3D model of the patient's head and deliver an appropriate diagnosis.

The visualization allows the doctor to make decisions based on what she sees. There is an underlying assumption at play: the visualization can be used for decision-making because it can be "trusted," i.e., it correctly represents the objects of interest in the patient's head. However, the visualization is clearly not an exact representation of all the complexities of the patient, but only a model (an approximation). It is important to understand what has been left out—i.e., what was ignored in the process—and what has been included—i.e., artifacts that do not represent any structure in the patient's head. Both can be understood by rigorously evaluating the decisions made at every stage of the pipeline: what kind of internal structures a CT machine can detect, how many images should be taken and at which radiation level, what algorithm should be used to reconstruct the 3D field, and which visualization method should be applied.[1] In other words, each decision is accompanied by error sources,[2] either intrinsic to the methodology used or introduced by external factors. As these errors sources creep in, the resultant image quality and reliability are affected. The diligent researcher is aware of these errors and tries to mitigate their effects.

---

[1]Even the choice of a CT machine as a way to investigate the patient's body already constrains the range of tissues that can be represented by the model. Because x-rays are used, only certain types of tissues will attenuate the x-rays enough to be perceptible, such as bones and skin; muscles, on the other hand, will not be represented. An MRI machine, for instance, has different constrains and can reveal other structures within the patient's body.

[2]Here the word "error" does not translate to "bad image," but any influence that causes the results to deviate from what the image should represent. A well-understood error source can be useful to the verification of visualization techniques.

For many decades now, other communities, such as computer science and computational science, have built tools to mitigate the diverse sources of errors. Examples of such tools include software testing (unit test, regression test, etc.), verification and validation, and version control systems. Visualization, on the other hand, needs a more tailored set of tools to make it more reliable. In the next section, we will review some of the tools often used in the visualization community to mitigate errors, thus building reliable visualizations. In this book, we will introduce techniques that help mitigate errors during the data mapping step of particular visualization techniques.

2.3   BUILDING RELIABLE VISUALIZATIONS

In our previous example, a visualization was used to help a doctor diagnose the cause of head pain. This is an illustration of how visualizations can be used in critical situations. In many fields of science, visualization is the lens through which scientists understand and evaluate their data [24]. The increased importance and widespread adoption of visualization tools entails *reliability*. As decisions are increasingly made by evaluating visualizations, the consequences of unreliable visualization range from wasted time caused by misleading results to unnecessary surgical procedures. Some of the errors that may creep in during the visualization pipeline include those due to acquisition problems (e.g., the patient moved during data collection), numerical truncation (the data was downsampled), implementation errors (the code has a bug), even algorithmic errors (the implementation is correct, but the underlying algorithm is wrong). There is no one method capable of dealing with all these errors. Throughout the years, each of these errors has been studied and dealt with to improve the quality of the overall pipeline, some more than others. As an example of the difficulty behind fixing these problems, we report next an historical account of the pursuit of a solution for the ambiguity problem of an Marching Cubes (MC) algorithm, a classic visualization algorithm.

2.3.1   THE PURSUIT OF A CORRECT MARCHING CUBES ALGORITHM

Medical equipment such as a CT scan or MRI provides a way of looking into the patient's body without surgical intervention. The results of using such equipment are not 3D models of organs but only raw data. In our previous example, the data is numbers representing the x-ray attenuation produced by a CT machine. The MC algorithm [30] is one of the classic techniques for mapping data to 3D virtual models with which users can interact as well as manipulate and evaluate.

The 3D virtual models are called *isosurface*. An isosurface I is the set of points for which a function f has a constant value c. One can think of isosurfaces as regions in space with similar properties. In our example, the isosurface I could represent the patient's skin tissue, bones, arteries, etc.; the similar property function f is the attenuation coefficient for each body part, i.e., how much an x-ray traversing through an artery, bone, skin, etc., is absorbed in the process; and the constant value c is the attenuation value associated with the region of interest (for instance, the skin instead of bones). Figure 2.5 shows two isovalues a user might be interested in: the first

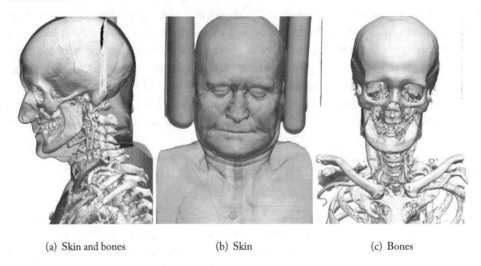

(a) Skin and bones   (b) Skin   (c) Bones

Figure 2.5: Isosurfaces corresponding to the skin and bones. (a) A side view of the skin and bones; (b) and (c) show the skin and bones, respectively. Both the skin and bones were extracted from the same scalar field, using the same method but with different isovalues.

is associated with the skin and the second with bones. Mathematically, an isosurface is defined as follows.

Definition 2.1   Let $\mathbf{x} \in \mathbb{R}^n$ be a point in $\mathbb{R}^n$, $c \in \mathbb{R}$ a constant and $f : \mathbb{R}^n \mapsto \mathbb{R}$ a real-valued function. The isosurface I, with corresponding isovalue c, is defined as:

$$I \;=\; \{\mathbf{x} \,|\, f(\mathbf{x}) = c\}. \tag{2.1}$$

In $\mathbb{R}^2$, I i is an *isoline*. The function f defines a height field, and the isolines are level-sets, or the regions in a map with constant elevation (see Figure 2.6). A hiker walking through a mountain trail that happens also to be an isoline will never go up or down the mountain but will walk at the same altitude throughout the trail.

Marching Cubes is one of the most important and widely used isosurface extraction techniques. It has an array of advantages that help make it popular, the first of which is *simplicity*. The MC algorithm can easily be explained and understood, both in 2D and 3D. The case table is straightforward and the method simple to implement. The MC algorithm is also *robust*. Regardless of the input grid, the algorithm will always stop and provide a surface as output. The MC algorithm is also *fast*, each voxel is visited only once, and the processing can be made in parallel. An enormous body of work has been published over the years, extending the MC in several directions. For a comprehensive discussion of MC-based isosurface techniques and the

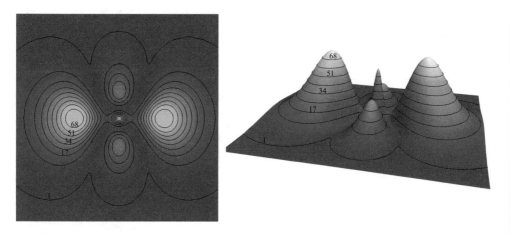

Figure 2.6: Isolines (black lines) of a scalar field (left). By mapping each point of the scalar field to altitude, we obtain a surface with bumps that resemble mountains (right). The isolines represent the points with the same altitude.

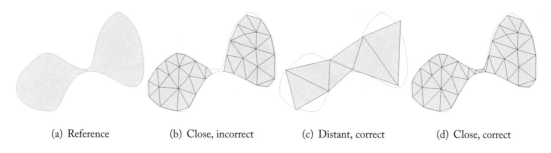

| (a) Reference | (b) Close, incorrect | (c) Distant, correct | (d) Close, correct |

Figure 2.7: (a) An isosurface. Two problems are shown: (b) is topologically incorrect, but geometrically close; and (c) is geometrically distant, but topologically correct. (d) The correct approximation of (a).

many improvements—geometry quality, parallel processing, view-dependent rendering, among others—we refer the interested reader to the survey by Newman and Yi [39]. We will discuss in detail the verification of isosurface extraction techniques in Chapter 4. For now, we will show some of the challenges researchers have faced throughout the years in the search for a correct MC algorithm.

Although MC has many advantages, at the time of its publication there were also problems. Ideally, surfaces extracted with any isosurface extraction technique should be both geometrically close and topologically equivalent to the isosurface I. The former states that points over the extracted surface mesh should be close to the isosurface I, whereas the latter states that the shapes of both M and I are the same. Figure 2.7 illustrates these concepts; see also Figure 2.8.

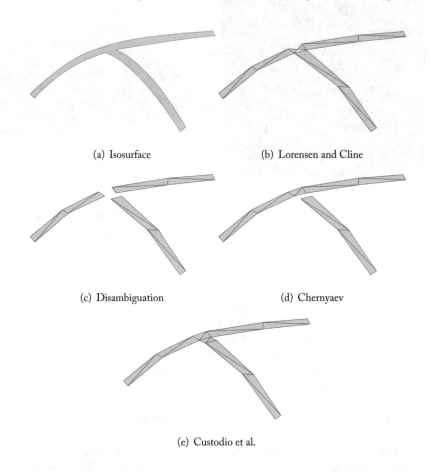

(a) Isosurface

(b) Lorensen and Cline

(c) Disambiguation

(d) Chernyaev

(e) Custodio et al.

Figure 2.8: Figure (a) shows the isosurface of interest. (b)–(e) Illustrates how MC-based techniques—used to extract a triangular mesh from isosurfaces—yield distinct results.

Here we describe the pursuit of an MC algorithm capable of preserving the shape of f. Suppose one wants to extract the isosurface illustrated in the Figure 2.8(a). Let us assume (a) to be a model of an artery. This case is challenging because it contains thin features that can be hard to represent. We will illustrate how the different MC-based techniques could fail with our hypothetical artery.

Soon after Lorensen and Cline's Marching Cube's article was published in 1987, its first problem surfaced [30]. In 1988, Dürst [8] described a problem with the polygonization of some pairs of neighboring cells. Because neighboring cells can be processed independently, surface cracks—i.e., holes that expose the interior of the surface—may appear. In our example, this is illustrated in Figure 2.8(b).

This problem was elegantly solved in 1991 by Nielson and Hamann's Asymptotic Decider [41]. The authors noted that the genesis of the problem is cells, which do not agree on the correct surface tiling. The proposed solution is simple: an unambiguous method to decide whether to join or separate the interior of the scalar field f. Nielson and Hamann proposed to use the sign of the critical point of the ambiguous face as the criterion to join or split the scalar field interior. In 1994, Montani et al. [37] published an alternative solution based on a consistent evaluation of the face signs. We illustrate a possible outcome in the Figure 2.8(c). Note that the resulting image no longer contain holes, but it fails to preserve the shape of (a).

These techniques ended the pursuit of a crack-free surface on MC. Nevertheless, in 1991 Natarajan [38] noted that even though these techniques produced crack-free surfaces, they did not preserve the shape of the isosurface, which is illustrated in our example by the fact that Figure 2.8(c) could have been connected in a different way. Another type of ambiguity is the interior ambiguity. In 1994, Natarajan looked at a critical point inside a grid cell to solve that problem; in the same year, Chernyaev published a new algorithm called MC 33 [4]. The name comes from the expansion of the MC lookup table to 33 cases (instead of 15) to cover all ambiguous topology.[3] Chernyaev also proposed a new procedure to select among the 33 cases.

Although the algorithms proposed by Chernyaev and Natarajan seemed to have solved the problem of interior ambiguity, it was found later that they do not always produce topologically correct triangulated surfaces. For instance, in 2003, Lopes and Brodlie [29] showed that Natarajan's work did not take into account the fact that some grid cubes may have two body saddles. As a result, some cases could be misclassified, leading to incorrect topology. We illustrate a possible outcome in Figure 2.8(d). Although the proposed algorithms preserve the topology for most cases, they can fail to correctly connect some features.

Lopes and Brodlie extended the Natarajan tests in order to correctly retrieve the topology of isosurfaces, but not all cases were included [39], which could still result in the wrong topology. In addition, the algorithm proposed is more complicated and, as far as we know, no implementation is available. Only in 2001 was a formal proof showing all possible cases of a topologically correct Marching Cubes published by Nielson [40]. Nielson also described a new algorithm for solving the interior ambiguity, but no implementation was provided.

Many of the techniques discussed thus far were not accompanied by an implementation that could be tested outside the laboratories in which they were developed. In 2003, Lewiner et al. implemented Chernyaev's technique and addressed some of its problems [27, 28]. Nevertheless, core problems were not addressed, thus sometimes leading to incorrect results. In 2013, Custodio et al. [5] uncovered problems in the work of both Chernyaev and Lewiner et al. Interestingly, both implementation and algorithmic problems were uncovered. The authors proposed further changes

---

[3]Interestingly, two of the 33 cases were redundant, which means that only 31 cases are needed [29]. The technique, nevertheless, is still called Marching Cubes 33.

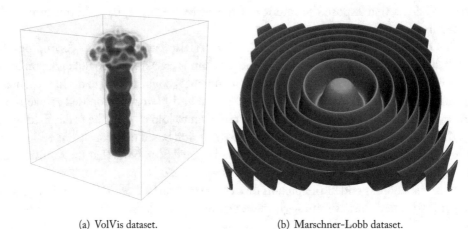

(a) VolVis dataset.   (b) Marschner-Lobb dataset.

Figure 2.9: Two examples of datasets used for verification purposes. (a) Fuel data, from the VolVis.org project. VolVis is a rich repository of volumetric data. (b) The Marschner-Lobb dataset is a widely used verification technique. Over 470 TVCG articles have referenced it.

to Chernyaev's algorithm and its implementation to achieve a topologically correct isosurface extraction technique.

This story highlights the difficulty in devising a provably correct algorithm and its correct implementation. Many of the pitfalls faced through the development of a topologically correct algorithm and implementation could be avoided by a systematic, thorough verification, as is done in other communities. We present some of the tools that can be used for verifying isosurface extraction algorithms in detail in Chapter 4. Before introducing these tools, we first review some of the practices already employed for verifying visualizations.

2.4   PATH TO VERIFICATION

The visualization community has been very active in the pursuit of techniques that increase the reliability of algorithms and implementations. Throughout the years, many steps have been taken toward building a culture of verification. For instance, Montani, Scateni, and Scopigno [37] proposed a technique for solving the previously explained MC ambiguity problem. The authors proposed a general solution, and thus, they needed to stress-test their code to demonstrate the reliability of the solution when dealing with the complex cases described in the literature. They used two approaches to verify their technique: (i) a function for which they knew the topology, so they could test whether the results are correct and accurate; and (ii) real-world datasets. Both solutions help increase code reliability, but as we will see in the next few chapters, sometimes it is not enough. Al Globus is one of the first researchers to formally acknowledge the need for verifying visualizations. Globus and Uselton [16] argue that visualization not only should be thoroughly verified, but the results of the verification process should be made available. Without some level of verification, visualizations are only "pretty pictures" [15]. The authors put it harshly:

> Other than blind faith, there is no reason to believe the results from visualization systems are more than approximately accurate most of the time.

Globus and Uselton use the term "evaluation" to encompass both human-centric evaluation and traditional verification and testing techniques. The authors note the need for many of the evaluation techniques we see today, such as *benchmark datasets* tailored to certain visualization techniques, *error characterization*, and *experiments with users*. Over the years, some benchmark datasets have been built and are now widely used. For instance, the VolVis project [35] is a rich source of volume data for testing volume visualization techniques. Another is the "Marschner-Lobb dataset," developed by Marschner and Lobb [32] and used to compare several reconstruction filters. This dataset (shown in Figure 2.9) has also become a standard in volume visualization. In the next chapter, we explain an alternative method for verifying isosurface extraction algorithms that can be used along with other techniques.

CHAPTER 3

Validation and Verification in Simulation Science

In this chapter, we seek to highlight some places one can recognize validation and verification (V&V) in the simulation science process, and to help glean from these examples some of the general principles of V&V. For those interested in the history of V&V in engineering, we point the interested reader to [42].

The use of simulation science as a means of scientific inquiry is increasing at a tremendous rate. The process of mathematically modeling physical phenomena, experimentally estimating important key modeling parameters, numerically approximating the solution of the mathematical model, and computationally solving the resulting algorithm has inundated the scientific and engineering worlds. In particular, the process has allowed for rapid advances in our understanding and utilization of the world around us. As more and more science and engineering practitioners advocate the use of computer simulation for the analysis and prediction of physical and biological phenomena, the computational science and engineering community has began to ask very introspective questions, such as the following [2].

- Can computer-based predictions be used as a reliable basis for making crucial decisions?

- How can one assess the accuracy or validity of a computer-based prediction?

- What confidence (or error measures) can be assigned to a computer-based prediction of a complex event?

In 2004, Patrick Roache [46] put forth the call to simulation scientists to take seriously the issue of V&V.

In an age of spreading pseudoscience and anti-rationalism, it behooves those of us who believe in the good of science and engineering to be above reproach whenever possible. Public confidence is further eroded with every error we make. As Robert Laughlin noted in this magazine, "there is a serious danger of this power [of simulations] being misused, either by accident or through deliberate deception." Our intellectual and moral traditions will be served well by conscientious attention to verification of codes, verification of calculations, and validation, including the attention given to building new codes or modifying existing codes with specific features that enable these activities.

In Roache's view, "conscientious attention" to these issues is necessary for simulation science to progress as a well-respected, actionable tool in the hands of those seeking public good. Based on Roache's quote and using [2] as our guide, let us first define verification and validation.

- **Verification** is the process of assessing software correctness and numerical accuracy of the solution to a given mathematical model.

- **Validation** is the process of assessing the physical accuracy of a mathematical model based on comparisons between computational results and experimental data.

Validation and verification (V&V) are thus considered the primary processes for assessing and quantifying the accuracy of computational results. From these definitions, we can make a few observations about how validation and verification interact and the focus of each process.

First, let us discuss the interaction of V&V. We have listed verification before validation because of the onion-shell nature of the V&V process. Most validation efforts use a computational (numerical) simulation code, and hence the entire validation process is in part predicated on proper verification having been done. Furthermore, the verification process can often be delineated into two nested components: algorithmic (numerical) verification and code verification. Given a model written in some mathematical form, the first stage in driving toward a solution is to make choices concerning the numerical approximations that will be used (and their corresponding algorithms). Once these choices have been made, one then must decide how to transform these high-level algorithms into executable code (through a choice of programming language, etc.). Only when one has gotten to the point of executable code does one obtain (in the best case) an actual "solution" to the problem of interest. How accurate that solution is depends on cascading effects: (1) how accurately was the algorithm transformed or transcribed into executable code (i.e., are there any coding bugs); (2) how good of an approximation does the chosen numerical algorithm provide; and lastly, assuming something that is bug-free and for which one have reasonable confidence in the numerical approximation, (3) how accurately does the model describe the physical phenomena of interest? The first two questions concern the areas of code verification and numerical (algorithmic) verification; the last question concerns the question of validation. Clearly, they build upon each other—requiring care at every step and tests that highlight possible deficiencies at every level.

As an example of this process, consider the following: imagine one were attempting to verify the correctness of a numerical implementation of Hookean dynamics—weight suspended by a spring as shown in Figure 3.1 (bottom left). The spring is stretched out, and if the string is overextended (pulling on the dangling object), we would find that the spring-mass system would oscillate. Natural damping in the spring will cause the oscillation to diminish in amplitude to the point that the spring is once again stretched out, balancing its internal forces with the load placed on it by the block.

Now imagine that we write down the ordinary differential equation system expressing the oscillator of the spring and solve it numerically. In Figure 3.1, we provide two graphs where a

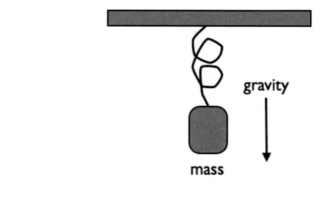

gravity

mass

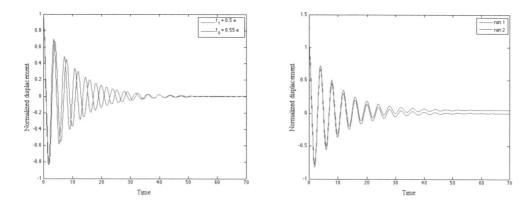

Figure 3.1: Hookean dynamics example: A diagram showing a block suspended by a spring (top). Displacement of the block if pulled down from its resting position (left). This figure is used to discuss numerical verification. The figure on the right is the same as the left but now with a coding bug introduced (right).

small mistake due to the choice of the numerical scheme and its parameters was made in the oscillatory frequency. The graph shows two examples of an oscillator, but which is correct? If we were interested only in the resting position, you might say you do not care as they give the same result. However, if we were interested in the particular dynamics of the problem, we would care. In this case, if we were to decrease the time step used in the numerical scheme, we would see that the red curve slowly "converges" to the blue curve—demonstrating to us that through the verification process we know that we must be careful about making any particular statement about the oscillating frequency without stating clearly what time integration scheme we used. In the second graph (Figure 3.1 (right)), we demonstrate a case where the numerical scheme is run with sufficient fidelity that the response amplitudes are the same, but a coding bug was introduced, which caused the final resting position to be slightly off. If only the two plots are shown, would we be able to tell what issues caused what? Maybe, in the rightmost plot, the numerical scheme approximated the tension coefficient for the spring incorrectly, yielding slightly different resting positions. Or maybe in the left plot, an indexing error caused a slight phase shift in the solutions that was eventually overcome by the damping nature of the problem.

3.1   A CANONICAL EXAMPLE

To help the reader understand components of the verification process, let us consider the following example. Consider that we need to compute the integral of a trigonometric function given by:

$$I = \int_0^{\pi/2} \cos(x)dx = \sin(\pi/2). \tag{3.1}$$

In this particular case, we know the exact solution, so we have something against which we can compare our answers. Imagine that the above equation is a mathematical model of some natural phenomena. As we said earlier, we are not at this stage testing the "validity" (i.e., validation) of this model, but rather want to figure out how to discretize it using numerical methods and ascertain how close we are to the answer. We begin by transforming our mathematical model into a numerical approximation (expression). In this case, let us consider using both *Riemann* integration and *trapezoidal* rule integration, given by:

$$R(h) = \sum_{i=0}^{N} h \cos(ih) \tag{3.2}$$

$$T(h) = \sum_{i=0}^{N} \frac{1}{2} h \left(\cos(ih) + \cos((i+1)h) \right), \tag{3.3}$$

where N denotes the number of intervals, i.e., $N = \pi/(2h)$. We define the error in the Riemann approximation as $E_R(h) = |I - R(h)|$ and the error in the trapezoidal approximation as $E_T(h) = |I - T(h)|$. From knowledge of the methods, we know that Riemann integration has *first-order* error characteristics—that is, if we decrease the spacing h that we use in our numerical approximation by half, we expect the error to go down by a factor of half. For trapezoidal rule integration, we expect second-order behavior: decreasing the spacing by half will lead to the error going down by a factor of four. These statements are the aforementioned "characteristics" or properties of the numerical approximation that we use as part of the verification process. When testing our implementation, we want to see if these properties of the schemes can be realized. If they are, we gain further confidence that our scheme has been properly implemented. If they do not, we need to figure out if we have incorrectly implemented something *or* if we have not met one of the assumptions upon which the numerical methods were built (such as assumed smoothness of the integrand, etc.).

Once we have decided the numerical schemes we will employ, we then implement them in computer code. In this case, we have provided the Matlab code.

```
% Code comparing Riemann and Trapezoidal Rule Integration
for i = 1:6,
    % set spacing
    dx = 0.5*pi/(2^(i+4));
    riem = 0;
    trpz = 0;
    for j=0:(2^(i+4)),
        riem = riem + dx*cos(dx*j);
        % Trapezoidal rule
        trpz = trpz + 0.5*dx*(cos(dx*j)+cos(dx*(j+1)));
    end
    spc(i) = dx;
    % error for Riemann integration
    err_riem(i) = abs(1.0-riem);
    % error for Trapezoidal rule integration
    err_trpz(i) = abs(1.0-trpz);
end
```

If we run the code given above and plot the results of the error against spacing on a log-log plot, we will obtain the results shown in Figure 3.2. The red circles denote the errors in the approximation using Riemann integration, and the blue circles denote the error using the trapezoidal rule.

This example serves to demonstrate two points. First, it shows us how we can use characteristics of the numerical methods we have chosen to give us some criterion against which to verify our code. Here we are able to state the numerical approximation as made concrete by our Matlab code demonstrates characteristic behavior, which does **not** mean that we can declare it to be bug-free (without error). Indeed, if you take the Matlab code above and change the j loop index to start from 1 instead of 0 – i.e., purposely insert a bug—you will get the same error plot for the range of spacing we have provided. Why? Verification is a process that requires a variety of tests (i.e., test harness) that in different ways "test the mettle" of the choices that were made and their implementations. This particular test, although useful in telling us that we have gotten the basic characteristics of the numerical methods right, does not have high enough fidelity (sensitivity to things such as loop indices) to tell us whether or not there is an indexing bug. Verification cannot be accomplished by a single magic test. Rather, it is a continual process of evaluating of our codes to see if they meet our expectations. If they do not, we need to inquire why.

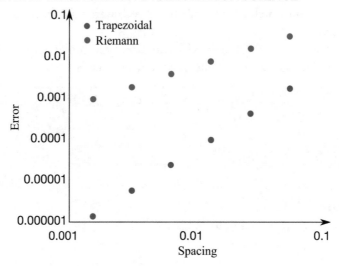

Figure 3.2: Error between our numerical approximation and the true solution versus mesh spacing. Red denotes Riemann integration and blue denotes trapezoidal rule integration.

3.2   A REALISTIC EXAMPLE

As a realistic example from the engineering literature, let us examine Kirby and Yosibash [25]. In this work, a well-known numerical method was applied to a new application. The application is called von-Karman plate dynamics, which is a set of equations expressing the motion of a thin plate under a load. A schematic diagram of the problem is given in Figure 3.3. In this chapter, we will summarize two studies that were contained in Kirby and Yosibash [25]—a spatial convergence study and a temporal convergence study—both of which were needed as verification steps before demonstrating new results.

Spatial Convergence
To verify that the numerical scheme proposed in Kirby and Yosibash [25] satisfied the boundary conditions and that we still obtain the spectral convergence expected of the Chebyshev-collocation method (the numerical method used in the article), we solved the linear bi-harmonic equation subjected to the boundary conditions, initial conditions, and a forcing function in such a way that an exact solution is available.

In Figure 3.4, we plot the discrete L_∞ error defined as

$$L_\infty = \max_{x_i, x_j} |u_{approx}(x_i, x_j) - u_{exact}(x_i, x_j)| \tag{3.4}$$

taken over the collocation point grid (x_i, x_j) vs. the number of points used per direction evaluated at the time $t = 1$. A time step of $\Delta t = 10^{-6}$ was used (so that given the second-order convergence in time our numerical integrator, we should expect time dicretization errors on the order of

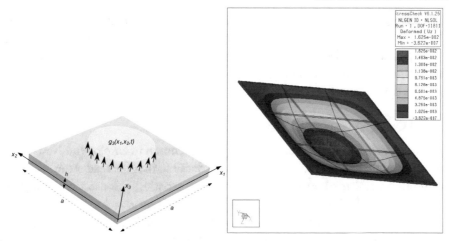

Figure 3.3: Schematic of the problem of interest: thin plate dynamics (left). An example of the displacement field obtained by placing the plate under a load with clamped boundary conditions (right).

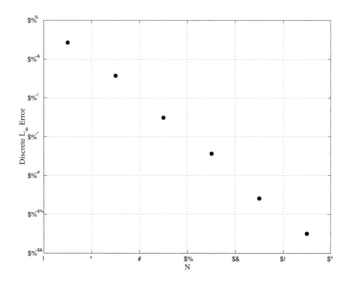

Figure 3.4: Discrete L_∞ error vs. the number of grid points used per direction in the Chebyshev collocation scheme. Data shown is taken at $t = 1$ with $\Delta t = 10^{-6}$.

10^{-12} and hence spatial errors should dominate); the exact solution was used to initialize the time integrator. Observe in Figure 3.4 that on a log-linear plot, a straight line is obtained, indicating an exponential convergence rate to the exact solution with increasing N.

This example serves to demonstrate the correctness of both the Chebyshev-collocation method and the boundary condition implementation used for a bi-harmonic PDE. As mentioned earlier, the verification process does not guarantee that the code is bug-free, but rather it shows that the numerical algorithm and its implementation meet the theoretical conditions established in the method development.

3.2.1   TEMPORAL DISCRETIZATION

To discretize the von-Karman system in time we have chosen to employ the Newmark-β scheme [21]. For this example, we will go into a little more detail so that the reader can see the connection between the numerical method choice and its characteristics. We use the average acceleration variant of the Newmark-β scheme (with Newmark parameters $\gamma = \frac{1}{2}$ and $\beta = \frac{1}{4}$), which exhibits second-order convergence in time, and is unconditionally stable under linear analysis.

The variant of the Newmark$-\beta$ scheme that we employed can be algorithmically described as follows. Assume one is given the equation:

$$m\ddot{u} + c\dot{u} + ku = g, \tag{3.5}$$

where the forcing g may be a function of the solution u. Discretizing in time we obtain the expressions at time level n and $n + 1$, respectively:

$$m\ddot{u}_n + c\dot{u}_n + ku_n = g_n$$
$$m\ddot{u}_{n+1} + c\dot{u}_{n+1} + ku_{n+1} = g_{n+1}. \tag{3.6}$$

The average acceleration variant of the Newmark-β scheme (see [21]) is given by the following time difference equations:

$$\dot{u}_{n+1} = \dot{u}_n + \frac{\Delta t}{2} (\ddot{u}_n + \ddot{u}_{n+1})$$
$$u_{n+1} = u_n + \Delta t\dot{u}_n + \frac{(\Delta t)^2}{4} (\ddot{u}_n + \ddot{u}_{n+1}), \tag{3.7}$$

where the local truncation error of these equations is $\mathcal{O}(\Delta t^2)$. We can now combine equations (3.6) and (3.7) to yield the following equation for the solution u at time level $n + 1$ given information at time level n and the forcing function g at time level $n + 1$:

$$\left(\frac{4m}{\Delta t^2} + \frac{2c}{\Delta t} + k \right) u_{n+1} =$$
$$g_{n+1} + m \left(\frac{4u_n}{\Delta t^2} + \frac{4\dot{u}_n}{\Delta t} + \ddot{u}_n \right) + c \left(\frac{2u_n}{\Delta t} + \dot{u}_n \right). \tag{3.8}$$

After substituting in the spatial discretization operators discussed previously, we arrive at a set of linear equations to solve. Given the linearity of the system, we can directly invert the

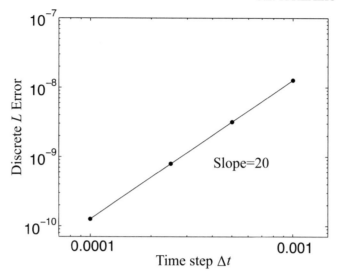

Figure 3.5: Discrete L_∞ error vs. the time step Δt. The comparison is done at time $t = 1$.

operator on the left-hand side to yield the solution u_{n+1} in terms of the solution u_n and the forcing $g(x_1, x_2, t)$ evaluated at the new time level t_{n+1}. In Figure 3.5, we plot the discrete L_∞ error versus the time step for a spatial discretization of $N = 21$ points per collocated direction. Observe that on a log-log plot, we obtain a straight line of slope 2.0, indicating that second-order convergence has been achieved.

CHAPTER 4

Isosurface Verification

In this chapter, we introduce the tools necessary for the verification of geometrical properties of isosurface extraction algorithms. As isosurfaces are ubiquitous tools used in scientific visualization, it makes sense to first understand how the ideas of error analysis and verification can be applied in the context of isosurfaces. We start by introducing the concept of isosurfaces and the problem of isosurface extraction. We then provide an overview of the verification technique, followed by a detailed description and application of the technique to real-world cases.

4.1 AN ISOSURFACE EXTRACTION PRIMER

4.1.1 MATHEMATICAL DEFINITION

A scalar field f' is a function that takes a point in an n-dimensional space and assigns a real value to it. A well-known example of a scalar field is a weather map. A temperature (real value) is assigned to each point of a map (2D space). Another example is the data generated by a CT scan. An attenuation (real value) is assigned to each point in the 3D space where the patient lies. Given a scalar field f', a *level set* is defined as follows.

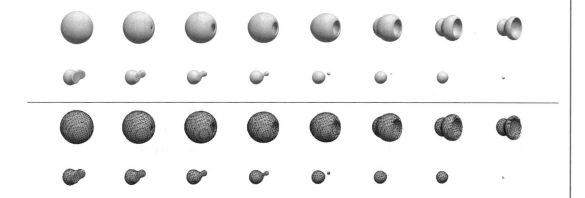

Figure 4.1: Top: as the isovalue increases, the set of points composing the isosurfaces changes accordingly. The isovalues of the isosurfaces above varied from 153–210. Notice how both the geometry and topology change as the isovalue increases. Bottom: approximation of the isosurfaces via triangulation.

Definition 4.1   Let $\mathbf{x} \in \mathbb{R}^n$ be a point in $\mathbb{R}^n$, $c \in \mathbb{R}$ a constant, and $f' : \mathbb{R}^n \mapsto \mathbb{R}$ a real-valued function. The level set I, with corresponding isovalue c, is defined as:

$$I \;=\; \{\mathbf{x} \,|\, f'(\mathbf{x}) = c\}.$$

Without loss of generality, we can assume $c = 0$:

$$\begin{aligned} I &= \{\mathbf{x} \,|\, f(\mathbf{x}) = f'(\mathbf{x}) - c = 0\} \\ &= \{\mathbf{x} \,|\, f(\mathbf{x}) = 0\}. \end{aligned} \tag{4.1}$$

Level sets are also written as $I = f'^{-1}(c) = f^{-1}(0)$, the preimage of f.

We are mainly interested in the case where $n = 3$, $f(\mathbf{x}) = f(x, y, z) = 0$, although higher dimensions are possible [58]. We will use the term *isosurface* to refer to 2-*manifold* level sets (see Figure 4.1). In a nutshell, a *manifold surface* is one in which the neighborhood of any of its points is *homeomorphic* to a (half-) disc. In other words, the neighborhood around a point looks like a (half-) Euclidean plane.[1] Next, we review one of the most commonly used technique for approximating isosurfaces using triangular meshes. The goal of this chapter is to introduce an algorithm for verify isosurface extraction techniques. How is an isosurface approximated? Typically, isosurfaces are approximated via triangulations.

4.1.2   ISOSURFACE APPROXIMATION

In the general case, isosurfaces cannot be represented exactly by a computer; thus, they are approximated. The scientific visualization literature is rich in methods for approximating isosurfaces. Here we are interested in approximations via *triangulated meshes* using the Marching Cubes (MC) algorithm. A triangulation is defined as follows.

Definition 4.2   Let $V = \{v_i\}$, $E = \{e_j\}$, and $F = \{t_k\}$ be a set of vertices, edges, and triangles, respectively. We say that the set $T = \{V, E, F\}$ is a triangulation if for any $a \in T$ and $b \in T$, the intersection $a \cap b = \emptyset$ or $a \cap b \in T$. A valid triangulation, e.g., ⋈ , should intersect only at edges and vertices. If the intersection between triangles does not occur at edges, e.g., ⋈ we say that T is not a triangulation, but a *triangle soup*.

The process of converting an isosurface into a triangulated mesh is referred to as *polygonization* or *isosurface extraction*. Many techniques can be used for polygonization of implicit surfaces; reviewing them is outside the scope of this chapter. We refer the interested reader to the book by Wenger [58] for a comprehensive review of isosurfaces. Here we review the extraction of surfaces based on the popular MC algorithm. MC is one of the simplest, fastest, and most robust techniques for isosurface extraction. We will illustrate how MC works in 2D space. Albeit its

---

[1]The (half-) disc condition is necessary in order to exclude degenerated cases, such as $f(x, y, z) = 0$, where $c = 0$ leads to a level set containing all points in $\mathbb{R}^3$. The neighborhood of any point of the level set is not a disc but a solid sphere. Another example is $f(x, y, z) = xyz$, where $c = 0$ leads to a level set that is the intersection of three planes. The neighborhood around the origin is not a single plane but the intersection of three planes.

implementation in the 3D case is more complicated, the ideas used in 2D extend naturally to the 3D case.

The isosurface extraction starts with the scalar field f. In practice, the scalar field f is not known at all $\mathbf{x}$ but only for a finite number of points in space. Typically, these points are located at the vertices of a rectilinear grid, shown in Figure 4.2. Hence, the first step of isosurface extraction algorithms is the interpolation of vertices values to produce a continuous scalar field. In 2D, a bilinear interplant is used. Its 3D counterpart is the trilinear interplant: $\bar{f}(x, y) = axy + bx + c$.

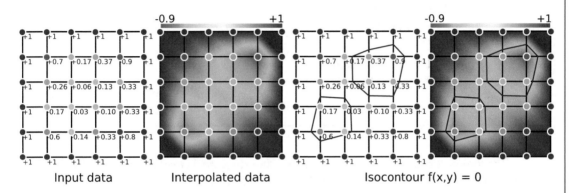

Input data   Interpolated data   Isocontour f(x,y) = 0

Figure 4.2: The MC 2D table cases. Only positive nodes (represented by the red dots) are shown for the sake of clarity. The left image shows all cases, and the right image shows the surface patches produced by each case.

The MC algorithm operates in each grid cell independently, checking whether the isocontour of interest intersects with that cell. Accounting for symmetry and rotation, there are only four types of intersections between the isocontour and a 2D cell, illustrated in the image on the right. The blue dots represent scalar values below the isovalue of interest c, whereas the red dots represent values above c. The isocontour starts at one grid edge that contains scalar values above and below c and progresses until

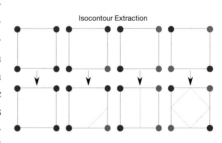

it reaches another edge of the same configuration. The bottom-right cell illustrates an ambiguous case in which both the continuous and the dashed isocontours are valid for that configuration. The ambiguity is solved by evaluating the sign of the critical point at the saddle point in that cell [41]. If the function value of the critical point is greater than c, then we separate the blue dots, as in ⊡ ; otherwise, we separate the red dots, as in ⊡ . This simple approach guarantees a consistent triangulation of ambiguous cases. These four cases represent the MC lookup table. For each cell, the MC algorithm must decide which case it belongs to and trace the isocontour within that cell.

Algorithm 1 The Marching Cubes algorithm.

MARCHINGCUBES(V, c)

    ▷ Let V be an input scalar field and c the isovalue of interest

1   **for** each pixel center (x, y) and ray direction **w**

2   **do** Add n samples $\{\mathbf{x}_i = \mathbf{x}(id)\}_1^n$ along **w**

3   **for** each $\mathbf{x}_i$

4   **do** $s_i = s(\mathbf{x_i})$

5               $\tau_i = \tau(s_i)$

6               $C_i = C(s_i)$

7               $I = I + (1 - \alpha)C_i$

8               $\alpha = \alpha + (1 - \alpha)\alpha_i$

9   **return** I

The lookup table and interpolation naturally extend to 3D volumes. Instead of only four cases, the table for the 3D cases contains 16 cases, some of which contain ambiguous configurations (see Figure 4.3). If not handled properly, these ambiguities can generate holes in the surfaces.[2] In this book, we will not be concerned with the table ambiguities, as they do not change the geometrical properties of isosurfaces. The algorithm for the 3D case is similar to that of the 2D. The result of the MC algorithm is a triangulation. Figure 4.4 shows an example of the extraction of the surface of an artery with an aneurism. The right image shows the typical MC triangulation pattern. Each of the triangles comes from a single voxel of the volume shown on the left.

4.2   OVERVIEW OF THE VERIFICATION PROCEDURE

Let $V_{N \times N \times N}$ be an $N \times N \times N$ rectilinear grid where the scalar field f is defined, c is an isovalue, and Iso is the MC implementation we wish to verify.[3] Our goal is to verify whether the triangulated surface $T_{N \times N \times N} = \mathrm{Iso}(V_{N \times N \times N}, c)$ is correct in the sense that its geometrical properties are representative of the actual isosurface $I = \{\mathbf{x} \mid f(\mathbf{x}) = c\}$. We will verify: (1) whether the function value at the centroid of each triangle $t \in T$ is a good approximation of the isovalue c; and (2) whether the normal $\vec{n}$ of each triangle vertex t is a good approximation of the gradient at the same point.[4] We note two aspects of the verification process, a practical aspect and a theoretical one. In this section, we focus on the practical aspects of the verification of isosurfaces.

---

[2]See Section 2.3.1 for an historical account of the pursuit of a correct MC.

[3]Although our main focus is on MC-like techniques, the verification tool presented here is by no means restricted to that class of techniques. In fact, the verification tool treats the implementation under verification as a blackbox, and thus any isosurface extraction technique can be verified. Nevertheless, the interpretation of the results changes depending on the algorithm used.

[4]Of course, other geometrical properties can be used in the verification process, e.g., curvature or area.

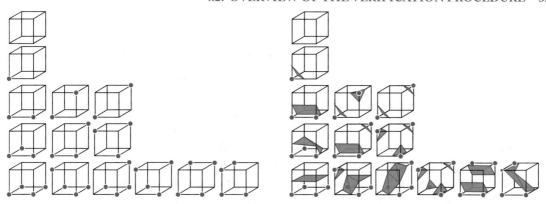

Figure 4.3: 3D cases for MC. Left: The data input consists of a rectilinear grid equipped with scalar values at each of its nodes. The blue dots represent points for which $f(\mathbf{x}) < 0$, and red dots are those with $f(\mathbf{x}) > 0$.

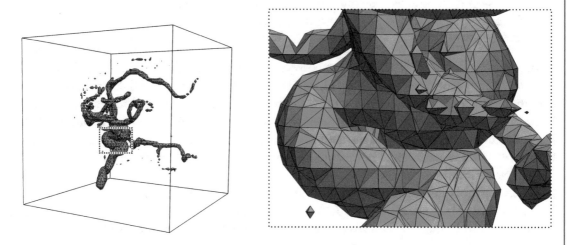

Figure 4.4: 3D mesh extracted using the MC algorithm. The box is the delimiter of the scalar field, which was obtained by a C-arm x-ray scan of a human head. The isosurface shows an artery of the right side of the head. The triangles extracted from the MC table are visible. The mesh corresponds to the isovalue $c = 40$.

The theoretical concepts described in this chapter are implemented within verifier , an open-source tool for verifying the correctness of isosurface extraction techniques. The first step is to install the verifier by typing the following into a shell environment:

```
$ curl -L http://tiagoetiene.github.io/verifier/install.sh | sh
```

The command shown above will download and install the verifier for isosurface extraction algorithms. `verifier` works by generating a set of grids (scalar fields) that will serve as input to the isosurface extraction technique under verification. `verifier` will then make a log file containing the results of the verification process. We start using it by simply typing:

```
$ verifier -N=4 --log=output.pdf <enter>
```

Once enter is hit, the `verifier` will never stop; instead, it will keep listening to any changes to the directory. To exit, just press `ctrl + c`. The output of the verification tool is V, a set of N volumes containing synthetic volume data with varying degrees of resolution. Mathematically, $V = \{V_{n_i \times n_i \times n_i}\}_{i=0}^{N}$. In the next step, the user runs the isosurface extraction technique under verification with the generated grids as input:

```
$ iso --grid=V_111 --output=T_111 [other options] <enter>
$ iso --grid=V_222 --output=T_222 [other options] <enter>
$ iso --grid=V_333 --output=T_333 [other options] <enter>
$ iso --grid=V_444 --output=T_444 [other options] <enter>
```

The output is a set T of N triangles meshes, each related to one of the volumes. Mathematically, $T = \{T_i | T_i = \mathrm{Iso}(V_{n_i \times n_i \times n_i}, c)\}_{i=0}^{N}$. Once T is available, `verifier` will detect it, read it, and output a document with the results of the verification procedure: a number k representing how fast the isosurface algorithm converges to the correct result and a plot illustrating the results of the test. k is a summary of the quality of the meshes produced by $\mathrm{Iso}(V, c)$. More specifically, `verifier` tests whether the vertices and normals of T_i converge to the correct position and normals of the isosurface I known to `verifier`. These ideas are illustrated in Figure 4.5. The idea behind the verification procedure is to detect whether the results of the isosurface extraction technique respect the properties of a *known* isosurface I. In the next section, we present in detail how geometrical properties can be used for isosurface extraction and how we evaluate the results. At the core of the `verifier` is the evaluation of the discretization errors of the triangular approximation of isosurfaces. The reader interested in the results of the application of the `verifier` to well-known isosurface extraction techniques should skip to Section 4.5.

4.3   DISCRETIZATION ERRORS

In order to verify isosurface extraction techniques, we must know *what to expect* from the output of the isosurface extraction technique under verification. We focus on how two properties of isosurfaces—algebraic distance and normal—are affected by its approximation via triangulation. The main results of this section are:

   i the algebraic distance of a triangulation converges *quadratically* to the isosurface I as a function of the voxel size h; and ii the triangulation normals converge *linearly* to the isosurface normal as a function of the voxel size h.

These two properties provide the expected behavior of MC-like implementations. The rest of this section is dedicated to the derivation of these properties. Readers interested in the implementation of the `verifier` should skip to the next section.

4.3.1 ALGEBRAIC DISTANCE CONVERGENCE ERRORS

The algebraic distance measures the difference between the function value of the triangulated surface and the isosurface I. The figure on the right illustrates the concept. As the voxel size diminishes, or equivalently, the grid is refined, the triangulation (shown in orange) better approximates the isosurface (shown in gray).

Note that the triangulation is only an approximation, and approximation errors cannot be completely eliminated. To see the errors, we colored the triangulation by using f (see Figure 4.6). Because the triangulation T represents an isosurface, we expect $f(\mathbf{x}) = 0$ for all $\mathbf{x} \in T$. By coloring all points of the triangulation (bottom row), we can see that this is not the case. The color of the function value at points close to the edge of the voxels closely resembles the color of the

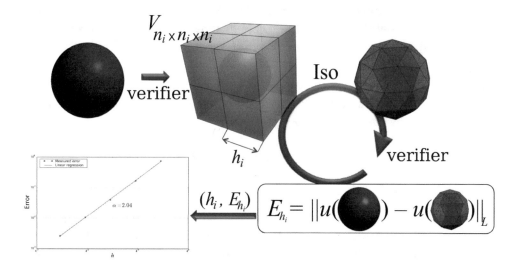

Figure 4.5: Verification pipeline. The pipeline starts with a known implicit function $f(x, y, z) = x^2 + y^2 + z^2 - 1$. We are interested in the isosurface $I = f^{-1}(c)$. The `verifier` first step is build volumes $V_{n_i \times n_i \times n_i}$. These volumes serve as input data to the isosurface technique under verification Iso. The output of Iso is a triangular mesh T_i, which is used to compute approximation errors between T_i and I. Last, the `verifier` generates a human readable report with the results of the verification procedure.

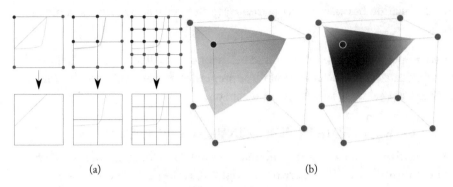

(a)   (b)

Figure 4.6: (a) A 2D example of error decay as the input grid is refined. (b) The error between the isosurface (left) and its approximation (right).

isosurface (gray). Nevertheless, as we navigate the triangulation toward the center of the triangles, errors are introduced, and the function value is no longer the same as the isosurface because, for those points, $f < 0$ (colored in blue). As the grid is refined, the effect is mitigated, but never eliminated. In order to verify isosurface extraction techniques, we must understand how fast the function value at the triangle converges to the isovalue of interest $f'(\mathbf{x_t}) \to c$, or equivalently $f(\mathbf{x_t}) \to 0$, as the voxel size $h \to 0$. The theorems shown in this section were extracted from Etiene et al. [13].

Theorem 4.3   *Let T be a triangulation approximating an isosurface $I = \{\mathbf{x} \in \mathbb{R}^3 | f(\mathbf{x}) = c\}$, c be an isovalue, and $\mathbf{x}_t \in T$ a vertex. The algebraic distance between T and the isosurface $f(\mathbf{x}) = c$ converges* quadratically *with respect to the grid size h:*

$$|f(\mathbf{x}_t) - \tilde{f}(\mathbf{x}_t)| = O(h^2). \tag{4.2}$$

Proof. In order to understand the errors introduced, we will use a simple Taylor expansion assuming linear interpolation along the grid edges. Let $C = [0, h]^3$ be a 3D grid cell and ijk its vertices. The function value is known only at the grid vertices, which we write as $f_{ijk} = f(\mathbf{x}_{ijk})$, and $\mathbf{x}_{ijk} = (x_i, y_j, z_k)$. Through a Taylor expansion of f, one can evaluate f at a point $\mathbf{x} \in C$ as:

$$f(\mathbf{x}) = f_{ijk} + \nabla f_{ijk} \cdot (\mathbf{x} - \mathbf{x}_{ijk}) + O(h^2), \tag{4.3}$$

where ∇f_{ijk} is the gradient of f in $\mathbf{x}_{ijk}$ and h the grid cell size. Let the linear approximation of f in $\mathbf{x}$ be defined by:

$$\tilde{f}(\mathbf{x}) = f_{ijk} + \nabla f_{ijk} \cdot (\mathbf{x} - \mathbf{x}_{ijk}) \tag{4.4}$$

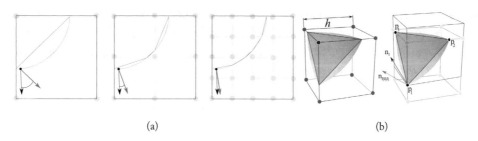

(a)   (b)

Figure 4.7: Normal convergence in (a) 2D and (b) 3D voxel. As the grid is refined, the approximation error descreases.

and consider a point $\mathbf{x}_t$ such that $\mathbf{x}_t$ is a point on the isosurface of $\tilde{f} = c$. The algebraic distance between the exact isosurface $f(\mathbf{x}) = c$ and the linearly approximated isosurface can be measured by $|f(\mathbf{x}_t) - \tilde{f}(\mathbf{x}_t)|$. From Equations 4.3 and 4.4:

$$\begin{aligned} |f(\mathbf{x}_t) - \tilde{f}(\mathbf{x}_t)| &= |f_{ijk} + \nabla f_{ijk} \cdot (\mathbf{x}_t - \mathbf{x}_{ijk}) + O(h^2) - \tilde{f}(\mathbf{x}_t)| \\ &= |\tilde{f}(\mathbf{x}_t) + O(h^2) - \tilde{f}(\mathbf{x}_t)| = O(h^2). \end{aligned} \tag{4.5}$$

Thus, the linearly approximated isosurface is of second-order accuracy.   □

4.3.2 NORMAL CONVERGENCE ERRORS

The rationale for normal convergence is similar to that of algebraic convergence, although the error analysis is more complex. Figure 4.7 illustrates the normal convergence as the input grid is refined. Intuitively, one expects the triangulated surface normals to converge to the isosurface normal as the grid size increases. In other words, the *angle* between the triangle normal $\mathbf{n}(x, y, z)$ and the isosurface normal $\nabla f(x, y, z)$ should diminish as the grid is refined. As before, we need to evaluate how fast the triangulation normals converge to the isosurface normal. The main result of this section, which will be used for verification purposes, is the following.

Theorem 4.4   *Let $\overline{\mathbf{n}}_t$ be the normalized normal vector of a triangle t (computed as the cross product of its edges) and $\overline{\mathbf{n}}(x, y, z)$ the normalized normal vector at a point $(x, y, z) \in t$. The dot product between $\overline{\mathbf{n}}_t$ and $\overline{\mathbf{n}}(x, y, z)$ converges linearly to 1 with respect to the grid size h:*

$$\overline{\mathbf{n}}_t \cdot \overline{\mathbf{n}}(x, y, z) = 1 + O(h).$$

The above theorem states that the angle between the triangle normal $\overline{\mathbf{n}}_t$, and the correct, known, normal $\overline{\mathbf{n}}$ goes to 0 as the cell size $h \to 0$, and it does so *linearly*. In order to prove this result, we first prove the following theorem.

Theorem 4.5   *Let T be a triangulation approximating an isosurface $I = \{\mathbf{x} \in \mathbb{R}^3 | f(\mathbf{x}) = c\}$ and $t \in T$ a triangle. The normalized normal vector $\bar{\mathbf{n}}_t$ of triangle t (computed as the cross product of its edges) converges* linearly *toward the normalized isosurface normal $\bar{\mathbf{n}}(x, y, z)$ at any point $(x, y, z) \in t$ with respect to the grid size h:*

$$\bar{\mathbf{n}}_t = \bar{\mathbf{n}}(x, y, z) + (O(h), O(h), O(h)) .$$

Proof. Our goal is to evaluate how normal vector errors accumulate. We assume that triangle normals are computed using the standard cross product between triangle edges. We will first derive the isosurface normal and then compare it with the approximated normal computed using the triangle mesh. Assume that the isosurface $f(x, y, z) = 0$ can be locally parameterized as $\Phi(u, v) = (u, v, g(u, v))$. Consider the triangle $t = \mathbf{p}_1\mathbf{p}_2\mathbf{p}_3$ defined by the points $\mathbf{p}_1, \mathbf{p}_2$, and $\mathbf{p}_3$ approximating the isosurface Φ in the grid cell. Without loss of generality, let us translate the grid cell so that one of its corners lies on $\mathbf{p}_1$, the origin of the coordinate system in which Φ is defined (see Figure 4.7), and so that the normals $\bar{\mathbf{n}}$ and $\bar{\mathbf{n}}_t$ are computed at $\mathbf{p}_1$. The normal $\mathbf{n}_1$ at $\mathbf{p}_1$ is:

$$\mathbf{n}_1 = \frac{\partial \Phi}{\partial u} \times \frac{\partial \Phi}{\partial v} = \left(-\frac{\partial g}{\partial u}, -\frac{\partial g}{\partial v}, 1 \right) . \tag{4.6}$$

Let $\mathbf{p}_1 = \Phi(u_1, v_1) = \Phi(0, 0)$, $\mathbf{p}_2 = \Phi(u_2, v_2) = \Phi(u_2, 0)$, and $\mathbf{p}_3 = \Phi(u_3, v_3) = \Phi(0, v_3)$, $u_2, v_3 > 0$. The triangle normal $\mathbf{n}_{\mathbf{p}_1\mathbf{p}_2\mathbf{p}_3}$ will be approximated using the cross product in $\mathbb{R}^3$:

$$
\begin{aligned}
\mathbf{n}_t &= (\mathbf{p}_2 - \mathbf{p}_1) \times (\mathbf{p}_3 - \mathbf{p}_1) \\
&= (u_2 - u_1, v_2 - v_1, g(u_2, v_2) - g(u_1, v_1)) \\
&\quad \times (u_3 - u_1, v_3 - v_1, g(u_3, v_3) - g(u_1, v_1)) \\
&= \begin{pmatrix} \mathbf{i} & \mathbf{j} & \mathbf{k} \\ u_2 - u_1 & v_2 - v_1 & g(u_2, v_2) - g(u_1, v_1) \\ u_3 - u_1 & v_3 - v_1 & g(u_3, v_3) - g(u_1, v_1) \end{pmatrix} \\
&= \begin{pmatrix} \mathbf{i} & \mathbf{j} & \mathbf{k} \\ u_2 & 0 & g(u_2, 0) \\ 0 & v_3 & g(0, v_3) \end{pmatrix} \\
&= (-v_3 g(u_2, 0), -u_2 g(0, v_3), u_2 v_3) . \tag{4.7}
\end{aligned}
$$

Equation 4.7 is the approximation obtained in practice. However, numerical errors are not explicitly represented. We solve this problem by first expanding $g(u, v)$ around (u_0, v_0). Let g_u, g_v, g_{uu}, g_{uv}, and g_{vv} be the partial derivatives of the first and second order of g at (u_0, v_0).

Then:

$$
\begin{aligned}
g(u,v) &= g(u_0, v_0) + g_u(u - u_0) + g_v(v - v_0) + \\
&\quad \frac{1}{2!}\left[(u - u_0)^2 g_{uu} + 2(u - u_0)(v - v_0)g_{uv} + (v - v_0)^2 g_{vv}\right] + \cdots \\
&= u g_u + v g_v + \frac{1}{2!}\left[u^2 g_{uu} + 2uv g_{uv} + v^2 g_{vv}\right] + \cdots \\
&= u g_u + v g_v + E(u, v).
\end{aligned}
\tag{4.8}
$$

The term $E(u, v)$ is the error involved in the approximation. Replacing Equation 4.8 into Equation 4.7:

$$
\begin{aligned}
\mathbf{n}_t &= (-v_3 g(u_2, 0), -u_2 g(0, v_3), u_2 v_3) \\
&= (-v_3 u_2 g_u - v_3 E(u_2, 0), -u_2 v_3 g_v - u_2 E(0, v_3), u_2 v_3) \\
&= u_2 v_3 \left(-g_u - \frac{E(u_2, 0)}{u_2}, -g_v - \frac{E(0, v_3)}{v_3}, 1\right) \\
&= u_2 v_3 \left((-g_u, -g_v, 1) + \left(\frac{O(h^2)}{O(h)}, \frac{O(h^2)}{O(h)}, 0\right)\right) \\
&= u_2 v_3 \left(\left(-\frac{\partial g}{\partial u}, -\frac{\partial g}{\partial v}, 1\right) + (O(h), O(h), 0)\right) \\
&= c\,(\mathbf{n}_1 + \mathbf{e}).
\end{aligned}
\tag{4.9}
$$

We now normalize $\mathbf{n}_t$. Given that $\|\mathbf{n}_1 + \mathbf{e}\|^{-1} = \|\mathbf{n}_1\|^{-1} + O(h)^5$ and $\|\mathbf{n}_1\|\mathbf{e} = \mathbf{e} = (O(h), O(h), 0)$:

$$
\begin{aligned}
\overline{\mathbf{n}}_t &= \frac{c\,(\mathbf{n}_1 + \mathbf{e})}{\|c\,(\mathbf{n}_1 + \mathbf{e})\|} \\
&= \frac{\mathbf{n}_1 + \mathbf{e}}{\|\mathbf{n}_1 + \mathbf{e}\|} \\
&= \frac{\mathbf{n}_1}{\|\mathbf{n}_1 + \mathbf{e}\|} + \frac{\mathbf{e}}{\|\mathbf{n}_1 + \mathbf{e}\|} \\
&= \frac{\mathbf{n}_1}{\|\mathbf{n}_1\|} + \mathbf{e}.
\end{aligned}
\tag{4.10}
$$

$\square$

Thus, we also need to derive the order of accuracy of the dot product between the two normal vectors. Next, we prove Theorem 4.4.

Proof. Recall that $\mathbf{e} = (O(h), O(h), O(h))$. Given that $\mathbf{e} \cdot \overline{\mathbf{v}} = O(h)$, where $\overline{\mathbf{v}}$ is a normalized vector:

$$\begin{aligned}
\overline{\mathbf{n}}_t \cdot \overline{\mathbf{n}}_1 &= (\overline{\mathbf{n}}_1 + \mathbf{e}) \cdot \overline{\mathbf{n}}_1 \\
&= \overline{\mathbf{n}}_1 \cdot \overline{\mathbf{n}}_1 + \mathbf{e} \cdot \overline{\mathbf{n}}_1 \\
&= 1 + O(h).
\end{aligned} \tag{4.11}$$

$\square$

4.4   VERIFICATION ALGORITHM

The derivation shown in the previous sections can be used to interpret the results of the verification process. Both algebraic distance and surface normal converge as a function of the grid cell size h. Thus, our verification procedure will progressively refine the input grid data and evaluate how errors decay. For algebraic distance, we evaluate errors as:

$$\begin{aligned}
E_i &= \max_{\mathbf{x}_t \in T} |f(\mathbf{x}_t) - \tilde{f}_i(\mathbf{x}_t)| \\
&= \max_{\mathbf{x}_t \in T} |O(h^2)| \\
&= \max_{\mathbf{x}_t \in T} |\beta_{\mathbf{x}_t} h_i^2 + \text{HOT}|,
\end{aligned} \tag{4.12}$$

where $\beta \in \mathbb{R}^+$, h_i is the grid cell size, and HOT are high-order terms. We assume that high-order terms can be neglected, and thus:

$$E_i = \max_{\mathbf{x}_t \in T} |\beta_{\mathbf{x}_t} h_i^2|. \tag{4.13}$$

---

[5]Let $||\mathbf{n}|| = O(1)$:

$$\begin{aligned}
||\mathbf{n} + \mathbf{e}||^{-1} &= \left\| \begin{matrix} a + O(h) \\ b + O(h) \\ c + O(h) \end{matrix} \right\|^{-1} \\
&= ((a + O(h))^2 + (b + O(h))^2 + (c + O(h))^2)^{-\frac{1}{2}} \\
&= (a^2 + b^2 + c^2 + O(h))^{-\frac{1}{2}} \\
&= (||\mathbf{n}||^2 + O(h))^{-\frac{1}{2}} \\
&= (1 + O(h))^{-\frac{1}{2}} / ||\mathbf{n}|| \\
&= \exp(-\frac{1}{2} \log(1 + O(h))) / ||\mathbf{n}|| \\
&= \exp(-\frac{1}{2}(O(h) - O(h^2))) / ||\mathbf{n}|| \\
&= \exp(O(h)) / ||\mathbf{n}|| \\
&= (1 + O(h)) / ||\mathbf{n}|| \\
&= ||\mathbf{n}||^{-1} + O(h).
\end{aligned}$$

The error analysis for normal convergence is similar. The pseudocode in Algorithm 2 shows how `verifier` can be implemented. In both cases, the error has the following format:

$$E_i = \beta h_i^k, \tag{4.14}$$

where k is the *observed convergence rate*. The observed rate is compared to the theoretical rate (quadratic for algebraic distance, $k = 2$, and linear for normals, $k = 1$). By taking the logarithm, k becomes the slope of the line connecting all observations E_i (see bottom image in Algorithm 2).

$$\log E_i = k \log(\beta h_i). \tag{4.15}$$

Algorithm 2 An algorithm for isosurface geometry verification.

$\text{VERIFY}(f, c, h_0, M)$

    $\triangleright$ Let $f : \mathbb{R}^3 \rightarrow \mathbb{R}$ be a scalar field and $c \in \mathbb{R}$ be an isovalue
    $\triangleright$ Let $I = \{\mathbf{x} \in \mathbb{R}^3 | f(x)\}$ be an isosurface
    $\triangleright$ Let h_0 be the initial grid cell size
    $\triangleright$ Let M be the number of meshes to be generated for testing convergence
    $\triangleright$ Let $\text{ISOSURFACEEXTRACTION}()$ be the technique under verification
1   $F_0 \leftarrow \text{ISOSURFACEEXTRACTION}(G, c)$
2   **for** $i \leftarrow 1$ **to** M
3   **do** $h_i = \frac{1}{2} h_{i-1}$
4      $T_i \leftarrow \text{ISOSURFACEEXTRACTION}(G_{h_i}, c)$
5      $E_i^{alg} = \max_{\mathbf{x}_t \in T_i} |f(\mathbf{x_t}) - c| \triangleright$ Algebraic distance errors
6      $E_i^{norm} = \max_{t \in T_i} |\nabla f(\mathbf{x_t}) \cdot \mathbf{n}_t| \triangleright$ Normal errors
7   Linear regression of $\{E_i\}_{i=1}^M$ using Equation (5.44):

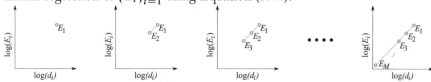

   The slope of the line is k, the expected convergence rate.
   The practitioner evaluates whether the observed behavior (the line slope) is close enough to the expected one ($k = 2$ for algebraic distance and $k = 1$ for normal convergence).

4.5 APPLICATION EXAMPLES

We apply the concepts illustrated in the previous section on two isosurface extraction implementations. The results presented in this section can be found in Etiene et al. [13]. We refer the interested reader to that article for a thorough evaluation including other geometrical properties and implementations.

VTK Marching Cubes   The first implementation we verify is VTK's implementation of the Marching Cubes algorithm [49]. Although there are other implementations available, we use VTK's implementation as it is a reliable library that has been thoroughly tested and has a sizable community behind it.

Macet: Macet [7] extends MC to improve the shape of the triangles in a mesh, thus improving mesh quality. Macet changes grid edges of the MC cases before their triangulation.

In order to compute the errors as shown in the previous section, we must define a scalar field f. We will use the following f in our tests:

$$f(\mathbf{x}) = f(x, y, z) = x^2 + y^2 + z^2 - 1, \tag{4.16}$$

and isovalue $c = 0$. The grid is defined in the domain $[-4, 4]^3$, with an initial grid cell size $h_1 = 1$ (an initial grid of resolution $8 \times 8 \times 8$). We progressively refine the grid cell by $h_{i+1} = h_i/2$, hence doubling the resolution. After each step of refinement, we resample the grid using f, instead of interpolating results, to avoid adding other error sources.

4.6   RESULTS

We applied the verification algorithm to the implementations shown in the previous sections.

4.6.1   VTK MARCHING CUBES

VTK's implementation of the MC algorithm obtained the best results in our tests. The results for both algebraic distance and normal tests were close to the theoretical result, namely $k = 1.94$ and $k = 0.93$, respectively. The figure shows the convergence curve for the algebraic distance test. Note that a positive result does not establish the implementation under verification as bug-free. The conclusion that can be drawn is conservative: "the test was not able to find issues that prevented convergence to the correct result." Typically, the verification procedure can reveal the presence of bugs but not the absence.

4.6.2   MACET

The expected convergence rate for algebraic convergence tests is $k = 2$. Our first results with Macet—shown in red—revealed that, after a certain number of refinements, the error remained constant. The plot shows a sharp error decline in the first refinement level, but then it slowly moves toward a fixed error value. The end result is a convergence rate of $k = 0.98$. This result suggests a source of errors other than the grid resolution becoming dominant after the first refinement.

The reason for the convergence problem lies in the way Macet solves the problem of poor triangle quality in one of the MC cases. To improve quality, Macet places an extra vertex *outside*

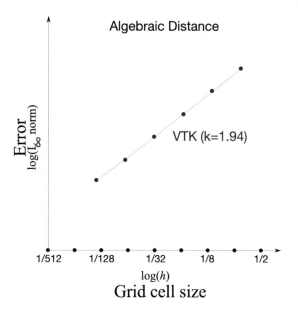

Figure 4.8: VTK's MC algebraic distance test. The expected and obtained curves are identical, which increases our confidence in the correctness of the algorithm.

the isosurface. Specifically, the implementation uses the center of the grid cell as one of the triangle vertices. Because the position of the new vertex is always at the center of the grid cell, the point often falls outside the isosurface and also changes the convergence plot. By forcing the inserted point onto the isosurface, the convergence curves change drastically (shown in blue). Note that, although the convergence rate is nearly zero, the numerical errors are low. The reason is that our implicit function is a simple distance field around a source point, and Macet uses a high-order spline interpolation which approximates the vertex position better than linear interpolation. Our theoretical analysis assumes that the isosurface extraction technique uses linear interpolation along edges in order to determine vertex position, not a high-order spline.

The normal convergence test also revealed the same problem. The expected convergence rate of $k = 1$ was not attained, and instead $k = -0.12$ was obtained. After fixing the problem just described, one obtains $k = 0.75$.[6]

4.7   DISCUSSION

A primary advantage of the verification procedure outlined in this chapter is that it is a nonintrusive method. The verification treats the implementation under verification as a blackbox, thus

---

[6]The latter result is much superior to the former, but it still begs the question of whether the result is close enough to the expected linear convergence rate. A more detailed investigation may reveal new sources of error responsible for "slowing down" the progress toward the correct normal vectors.

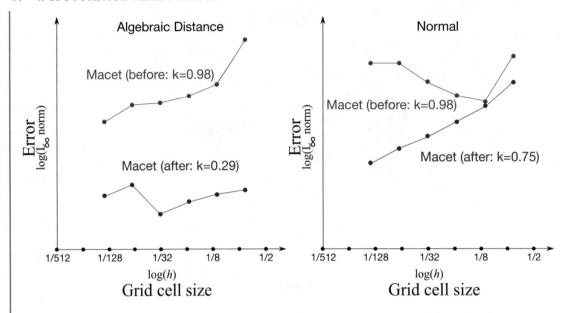

Figure 4.9: Macet's normal and algebraic distance test. The red curves show the convergence results before fixing an issue with the implementation. The blue curves show the result after the issue was fixed.

allowing easy integration with existing verification pipelines. On the other hand, the verification procedure does not provide clues as to where a bug may be hidden. In fact, errors in the verification procedure itself translate into a flawed convergence analysis. Therefore, when a mismatch between expected and obtained results occurs, the practitioner must carefully evaluate not only the code under verification but also the steps involved in obtaining the results.

A second important issue is the choice of the scalar field that is used as a manufactured solution. For the sake of simplicity, we have chosen a simple distance function. Nevertheless, development pipelines may benefit from more advanced functions. The choice of the scalar field is an important one. The simpler the scalar field, the less likely it is to reveal any potential issues with the isosurface extraction technique.[7] The results of the verification algorithm *cannot* determine the absence of problems. Instead, the results increase our confidence in the implementation under verification.

The hardest part of the verification procedure is the derivation of the expected (theoretical) behavior of algorithms. In Section 4.3, we derived the approximation errors for the algebraic distance and normal approximations between the isosurface $f = c$ and its MC-based triangula-

---

[7]In fact, we did observe this limitation in practice. We used the manufactured solution given by $f(x, y, z) = x + y$ to verify an isosurface extraction implementation known to have bugs, and the results were successful. Other scalar fields, on the other hand, revealed the known problem.

tion. These results can readily be used by anyone interested in verifying isosurface algorithms. The verification procedure can be further improved by testing other properties of isosurfaces, such as curvature, area, or volume. In this case, one must derive the theoretical analysis for that property of interest, which can be quite complex. Nevertheless, suppose that one is interested in knowing whether a given property converges to any solution as the grid is refined. In that case, one can build the convergence plot to see whether the implementation being verified produces results that converge toward the solution or not. By plotting convergence in this way, extra care must be taken because the code under verification may contain a bug and still converge to the correct solution.[8]

Besides the challenge of deriving the theoretical analysis of different isosurface properties, there is also the challenge of other isosurface extraction techniques. An example is advancing front techniques. These techniques typically build a triangulation by adding triangles on top of the isosurface without a lookup table. These techniques define (semiautomatically) a suitable triangle size when building a triangulation. In this case, the approximation may depend not only on the grid size but also on other properties, such as the triangle size. The theoretical analysis of these algorithms must consider errors introduced not by the grid size but by the triangle size.

Lastly, as we briefly described in Chapter 2, two aspects of isosurfaces should be considered when discussing about isosurfaces: geometry and topology. In this chapter, we have dealt only with the former. The latter is also important and a deeper investigation is beyond this introduction to the verification of isosurfaces, but it is detailed elsewhere [12].

4.8   CONCLUSION

The role of isosurface extraction techniques in the scientific pipeline has grown beyond merely generating pretty pictures. It is now fundamental to scientific inquiry. Hence, these techniques should be thoroughly verified. The technique presented in this chapter is an important step toward reliable isosurface extraction techniques. A number of steps can further increase the trustworthiness of the verification algorithm and, consequently, build a reliable isosurface extraction technique.

The main ingredients of the verification procedure are the *manufactured solution* and the *theoretical results* describing the behavior of the isosurface techniques. A public database of manufactured solutions would greatly help developers in the process of verifying the results of isosurface extraction techniques. Such a database would lift a hefty burden from their shoulders as they would not have to develop manufactured solutions—which should be robust enough to verify an implementation—from scratch.

In addition, a public database of theoretical results can be very useful not only for verification purposes but also to inform users of what to expect from isosurface extraction techniques. Such a database provides theoretical guarantees about the algorithms, not implementations, and

---

[8]Some code mistakes will affect the *rate* of convergence but not the convergence per se. An example is a high-order interpolation that uses wrong sampling positions (node- vs. cell-centered). The method will converge, but not at higher rates, as expected from a high-order method, which means that the extra computational effort spent on the high-order interpolant is lost because the convergence rate is mainly affected by the mistake introduced in the sampling position.

is composed of a set of (i) the geometrical property of interest, (ii) the approximation method, (iii) the isosurface extraction algorithm used to implement it, and (iv) the expected order of accuracy. For instance, what is the expected convergence of the (i) Gaussian curvature computed using the (ii) angle deficit method, of a triangulation built using (iii) Afront (an advancing front technique)? A database containing multiple instances of these three items for a variety of isosurface extraction technique can help users find the appropriate errors for each technique being used. Also, a public database of verified code is a valuable source for those interested in using only verified isosurface extraction techniques.

Along with the theory explained in this chapter, we provide a code that can be used to verify isosurface extraction techniques. The code is available at `http://tiagoetiene.github.io/verifier/` along with its documentation and examples. In the next chapter, we discuss the verification of volume rendering techniques, another method for visualizing volumetric data.

CHAPTER 5

Volume Rendering Verification

In this chapter, we introduce a technique for the verification of volume rendering algorithms. We start with a description of the volume rendering algorithm, and then discuss the the standard discretization procedures, and finally, how this information can be used to verify the algorithm correctness. At the core of the verification procedure, we use the order of accuracy and error convergence analysis. We show how a simple black-box algorithm can be used to verify volume rendering techniques.

5.1   A VOLUME RENDERING PRIMER

Direct volume rendering comprises a set of techniques developed for the visualization of volume data, e.g., data acquired from CT machines. The term "direct" is used to differentiate this type of volume rendering from other volumetric techniques that require intermediate geometry to visualize the data, such as isosurfaces (see Chapter 4). Direct volume rendering techniques— henceforth simply *volume rendering techniques*—act upon the volume data and generate an image, rather than a geometrical surface that needs to be further rasterized.[1]

The main idea behind volume rendering algorithms is to mimic the behavior of a ray of light going through a particle cloud. The interaction between the ray of light and the suspended particles in the environment can be modeled by a mathematical equation known as the *volume rendering integral (VRI)*. The solution of the VRI depends on a variety of factors, such as input data and numerical accuracy, and many of the algorithms available in the literature account for these factors. In this section, we provide the intuition behind the interactions but omit the equations derived from them that lead to the volume rendering equation. Instead, we refer the interested reader to the excellent work of Nelson Max on optical models for direct volume rendering for a detailed description and derivations [33].

In the simplest ray-particle interaction, the light traverses the particle cloud and part of its energy is *absorbed* by the particles in its way (Figure 5.1(a)), which happens because particles are not perfectly transparent. The equation below shows the volume rendering equation of an absorption-only model:

$$T = \exp\left(-\int_0^D \tau(s(\mathbf{x}(\lambda')))d\lambda'\right), \tag{5.1}$$

---

[1]Volume rendering does not replace isosurfaces and vice versa. Volume rendering techniques can better evaluate surfaces that do not contain a clear separation boundary, such as a cloud. Isosurface techniques, on the other hand, provide geometrical information that can be used to infer surface properties such as area, volume, and curvature.

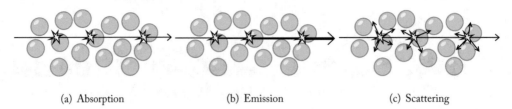

(a) Absorption   (b) Emission   (c) Scattering

Figure 5.1: Ray-particle interaction modes. Absorption: the incoming ray hits a suspended particle and part of its energy is lost. Emission: the incoming ray hits a glowing suspended particle and more energy is transmitted along with it. Scattering: the incoming ray hits a suspended particle and it is scattered in many directions.

where the solution T of the integral is the transparency of the medium, D is the ray length, $\tau(s)$ is the light extinction coefficient, and $s(\mathbf{x}(\lambda))$ is the scalar value at position $\mathbf{x}$ in the ray parameterized by λ. Another possible ray-particle interaction involves the *emission* of light as the ray traverses the particle cloud (Figure 5.1(b)). Each of the particles may have enough energy to emit light that adds up to the incoming ray. The equation derived from this scenario is:

$$I = \int_0^D C(s(\mathbf{x}(\lambda)))\tau(s(\mathbf{x}(\lambda)))d\lambda, \qquad (5.2)$$

where $C(s(\mathbf{x}(\lambda)))$ is the emitted light and the remaining terms are the same as before. In nature, both cases occur simultaneously and at different degrees, and thus they are used as the basis of many volume rendering algorithms available in the literature. More specifically, the *emission-absorption* model is commonly referred to as the VRI. In this chapter, we will consider only the verification of volume rendering algorithms implementing the emission-absorption model:

$$I = \int_0^D C(s(\mathbf{x}(\lambda)))\tau(s(\mathbf{x}(\lambda))) \times \exp\left(-\int_0^\lambda \tau(s(\mathbf{x}(\lambda')))d\lambda'\right) d\lambda. \qquad (5.3)$$

The previous equation has terms from both emission and absorption models. Another ray-particle interaction takes into account *scattering*. As the ray hits a particle, it may be not only absorbed, but also *scattered* in multiple directions (Figure 5.1(c)). Modern volume rendering engines do take scattering into account, but for simplicity we will consider only the VRI described in Equation (5.3). See Max [33] for more details.

5.1.1   SOLVING THE VOLUME RENDERING EQUATION

Equation (5.3) can be solved analytically only in special cases. We will use analytical solutions in Section 5.4 as part of the algorithm that verifies volume rendering techniques. Nonetheless, in the general case, Equation (5.4) must be solved numerically. A well-known approach for solving the VRI is *via volume ray casting* (see Figure 5.2). In volume ray casting, a ray starting at a

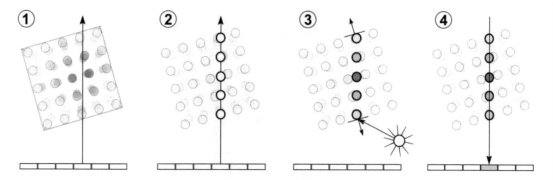

Figure 5.2: Four steps of the volume ray casting algorithm: (1) ray casting, (2) ray sampling, (3) shading, and (4) compositing. Source: Wikimedia Commons [56].

pixel position directed toward the volume is discretized into sample points; the sample points are then evaluated according to transfer functions, and, lastly, combined into one value via numerical integration. The typical discretization of Equation (5.4) leads to

$$ I \;=\; \sum_{i=0}^{n-1} C(s(\mathbf{x}(id)))\tau(s(\mathbf{x}(id)))d \left(\prod_{j=0}^{i} 1 - \tau(s(\mathbf{x}(jd)))d \right). \tag{5.4}$$

Sample points are defined at position $\mathbf{x}(id)$, where i is the i-th sample in the ray, and $d = D/n$, where D is the length of the ray and n is the total number of sample points. Now, let us explain in detail how each term on the right side of the equation is defined. $\mathbf{x}$ is the position of a point within a ray, and it can be easily computed for any ray. s is the input scalar field, e.g., data from a CT-machine. Finally, τ has to be implicitly defined by the user as a *transfer function*.[2] The user will also need to provide a relation between color and s. The intuition is the following: we expect certain scalar values to be associated with a structure that we are interested in. For instance, the scalar value s' could be associated with skin tissue. Thus, a transfer function C can be defined as a piecewise-linear function that is transparent for all scalar values $s \neq s'$ and reddish at $s = s'$ (see Figure 5.3). Algorithm 3 illustrates a volume rendering implementation.

5.2   WHY VERIFY VOLUME RENDERING TECHNIQUES

Volume rendering techniques can be used in critical situations, such as in medical diagnosis. It is important to make sure that the image seen on a screen can be used as a reliable basis for decision-making, otherwise there may be undesirable consequences. Consider the following documented example, the case of a false stenosis. A patient went through an unnecessary surgery because the

---

[2]We say *implicitly* because the user does not define τ as a transfer function but T directly.

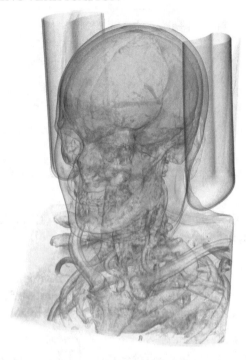

Figure 5.3: Transfer functions can reveal structures of interest. In this example, a transfer function was set to reveal detail of both of the skin and the bones.

images used to support diagnosis were misleading. The images revealed the presence of a stenosis that did not exist in reality, a false stenosis [31].[3] This case illustrates the importance of minimizing all source of errors from the visualization pipeline to help produce trustworthy images.

The volume rendering community has developed many techniques for dealing with problems of hidden data features, uncertainty of the input data, transfer function design, and others, to help users to maximize their chances of producing trustworthy images. We are interested in one unexplored area of techniques developed to improve the quality of the implementation of these

---

[3]The problem of false stenosis turned out to be in the transfer function, which was hiding important details of vessels from the rendered image. We can speculate what could have happened if the results of the transfer function were different. Let us assume that a faulty transfer function introduced a different problem in the image, for instance salt-and-pepper noise (black and white dots throughout the image). In this case, the problem with the transfer function is straightforward to detect because the noise is clearly not part of any internal structure of the human body. In other words, the resulting image is very different from our expectations. Unfortunately, in the case of false stenosis, the rendered image contained structures compatible with a known medical condition, which makes the problem harder to detect.

Algorithm 3 A simple algorithm for direct volume rendering via ray casting.

VOLUME RENDERING(s, τ, C)

$\quad \triangleright$ Let $\alpha_i \approx 1 - e^{-\tau_i d} \approx \tau_i d$ be the opacity at i-th sample
$\quad \triangleright$ Let I be the solution of the VRI
1 **for** each pixel center (x, y) and ray direction $\mathbf{w}$
2 $\quad$ **do** Add n samples $\{\mathbf{x}_i = \mathbf{x}(id)\}_1^n$ along $\mathbf{w}$
3 $\quad\quad$ **for** each $\mathbf{x}_i$
4 $\quad\quad\quad$ **do** $s_i = s(\mathbf{x_i})$
5 $\quad\quad\quad\quad$ $\tau_i = \tau(s_i)$
6 $\quad\quad\quad\quad$ $C_i = C(s_i)$
7 $\quad\quad\quad\quad$ $I = I + (1 - \alpha)C_i$
8 $\quad\quad\quad\quad$ $\alpha = \alpha + (1 - \alpha)\alpha_i$
9 **return** I

---

techniques, i.e., techniques designed to verify volume rendering techniques. We define volume rendering verification as in Definition 5.1.

Definition 5.1　A volume rendering technique (algorithm and implementation) is said to be verified if it correctly solves the VRI.

Definition 5.1 is straightforward and powerful. Note that correctly solving the volume rendering equation does not necessarily imply exact results, but results that could be made exact provided that arbitrary amounts of resources are available.[4] Because it is impossible in the general case to produce exact results, the verification will be completed when we increase our confidence that the results can be arbitrarily close to the exact solution. Using verification in this sense, the visualization community typically relies on two approaches for verifying volume rendering techniques: *expert analysis* and *error quantification*.

Expert analysis relies on the human visual system to detect anomalies in the image. It is a good test[5] and it can uncover many problems during the development cycle as it quickly reveals large-scale problems. Nevertheless, consider the image shown in Figure 5.4(a), which was generated with a buggy volume rendering implementation. The bug was introduced manually to evaluate its effect on the final image [11]. It is hard to detect any problem with that image, as it

---

[4]For instance, infinite precision arithmetic, unlimited CPU power to solve the VRI with an infinite number of sample points, and so on.

[5]Many everyday algorithms, e.g., quicksort, do not naturally result in an image or plot that can be visualized, but instead produce other kinds of output. One way of debugging these algorithms is by visualizing their results to understand what went wrong during their development. For instance, plotting the output of a sort algorithm may help reveal problems in parts of the implementation. Nevertheless, the latter step has to be done separately. In this sense, volume rendering developers are fortunate as the result of their work is an actual image that can be used to debug the algorithm being implemented.

seems perfectly valid, i.e., no undesirable artifacts are present. Nonetheless, the image is certainly not accurate from a numerical standpoint.

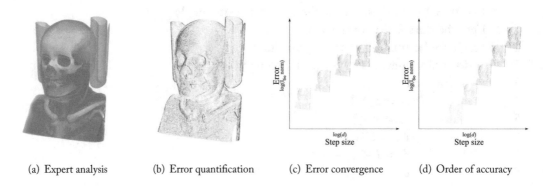

(a) Expert analysis   (b) Error quantification   (c) Error convergence   (d) Order of accuracy

Figure 5.4: Several techniques can be used to verify code. The degree of rigor of the techniques increases from left to right.

A second technique used for verifying volume rendering relies on an analysis of the numerical errors of volume rendering techniques. In this case, the volume rendering developer starts with a *ground truth*—a trustworthy image against which the results of the software will be compared. Next, the developer generates images that can be compared against the reference image. By doing this, the user can measure errors in the whole image and evaluate whether the error is acceptable (see Figure 5.4(b)). As in the case of expert analysis, in this case the user is also invited to make a judgment about what is considered acceptable. Error quantification is more rigorous than expert evaluation in the sense that it can drive the user's attention to regions of high error that may raise a flag about potential problems during development. Nevertheless, a problem remains. Is an absolute error of 0.01 small enough to consider the technique verified? Maybe 0.005?

Both expert analysis and error quantification are straightforward to apply, but they are not the most rigorous techniques that can be applied to verify volume rendering. In fact, two improvements can make error quantification more sensitive to programming and algorithmic mistakes. The first is *convergence analysis*. In convergence analysis, the user does not track errors in a single image, but errors from successive refinement of some of the discretization parameters, for instance, the ray sampling. Suppose that the initial size between ray samples is defined to be d. The user will then quantify the errors between a reference image and the rendered image, which will result in an error. Let us pick the maximum error among all pixels to be the error representative for the whole image. Now, let us volume-render another image, but this time using a sample distance of $d/2$. Intuition tells us that, with more samples, we will be able to obtain better images, which translates into lower error. By plotting the errors associated with each image, we can expect the errors to decrease as the model is refined. This process, illustrated in Figure 5.4(c), is, in a nutshell, convergence analysis. Convergence analysis uses successive error quantification to evaluate

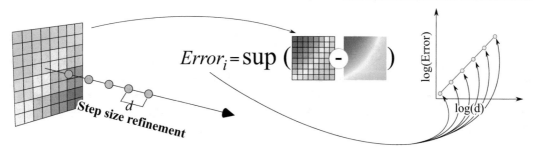

Figure 5.5: The verification pipeline of volume rendering techniques. The rendered images are generated by the progressive refinement of the space d between sample points (left). Next, the maximum error between the rendered images and a ground truth image is computed (middle). Finally, the error is plotted and the convergence rate is evaluated (right).

whether errors are decreasing as expected. The tricky part is to understand what the "expected" error decay is, which we will explain in detail in Section 5.4. Last, the order or accuracy method, shown in Figure 5.4(d), is a refinement of convergence analysis that checks not only whether the errors decay as $d \to 0$, but also the *speed* of decaying. The latter is considered to be the most rigorous method for verification within the computation science and engineering community.

Next, we explore the volume rendering equation and its typical discretization. The goal is to check what the expected behavior is as one increases the number of samples per ray progressively. We want to know not only whether the equations converge but also the speed of convergence.

5.2.1 OVERVIEW OF THE VERIFICATION PROCEDURE

The verification procedure is composed of four parts: (i) theoretical evaluation of the discretization errors introduced; (ii) a ground truth, which in our case is either an analytical solution of the VRI or the rendered images themselves; (iii) maximum error computation between the reference image and the rendered images; and (iv) plotting the errors and evaluating whether they correspond to the errors obtained in (i). Figure 5.5 illustrates the verification pipeline.

5.3 DISCRETIZATION ERRORS

The first step in the verification of the volume rendering algorithm is to describe how the discretization errors should behave, i.e., we must know not only whether the discretization used in Equation (5.4) converges to the correct solution, but also the speed of convergence. The VRI is discretized using a first-order approximation of the inner and outer integral and a second-order approximation of the exponential term:

$$\int_0^D f(s(\mathbf{x}(\lambda)))\mathrm{d}\lambda = \sum_{i=0}^{n-1} f(s(\mathbf{x}(id)))d + O(d), \tag{5.5}$$

and

$$e^x = 1 + x + O(x^2). \tag{5.6}$$

We will plug Equations (5.5) and (5.6) into Equation (5.3) and evaluate how the error spreads during discretization. In the next section, we present the details of the derivation of the linear convergence. In practice, the derivation has to be done only once, and the result can be used later for verification purposes. Readers interested in the verification algorithm can skip to Section 5.4. The derivation shown next was extracted from Etiene et al. [11].

Approximation of the Inner Integral $T(\lambda)$

Let $T(\lambda) = T_\lambda = e^{-t(\lambda)}$, where $t(\lambda) = \int_0^\lambda \tau(\lambda')d\lambda'$, and λ parameterizes a ray position. We will first approximate $t(\lambda)$ and then $T(\lambda)$. In the following, $d = D/n$ is the ray sampling distance, D is the ray length, and n is the number of subintervals along the ray:

$$\int_0^\lambda \tau(\lambda')d\lambda' = \sum_{j=0}^{i-1} \tau(jd)d + O(d), \tag{5.7}$$

where $\lambda = id$. Using Equation (5.7):

$$T(\lambda) = \exp\left(-\int_0^\lambda \tau(\lambda')d\lambda'\right) \tag{5.8}$$

$$= \exp\left(-\sum_{j=0}^{i-1} \tau(jd)d + O(d)\right) \tag{5.9}$$

$$= \left(\prod_{j=0}^{i-1} \exp\left(-\tau(jd)d\right)\right) \exp\left(O(d)\right). \tag{5.10}$$

Let us define $\tau_j = \tau(jd)$. We start with a Taylor expansion of $\exp(O(d))$:

$$T_\lambda = \left(\prod_{j=0}^{i-1} \exp\left(-\tau_j d\right)\right)(1 + O(d)) \tag{5.11}$$

$$= \prod_{j=0}^{i-1} \exp\left(-\tau_j d\right) + \prod_{j=0}^{i-1} \exp\left(-\tau_j d\right) O(d). \tag{5.12}$$

The second term in the right-hand side of Equation (5.12) contains only approximation errors (see the multiplying $O(d)$ term). Thus, we will expand it using a first-order Taylor approximation and use the fact that $\tau_j d = O(d)$:

$$\prod_{j=0}^{i-1} (1 - O(d)) O(d) = (1 + O(d))^i O(d), \tag{5.13}$$

where the change in the sign is warranted because the goal is to determine the asymptotic behavior. For $i = 1$, only one step is necessary for computing the VRI along the ray, and the previous equation will exhibit linear convergence. Nevertheless, in the general case, the numerical integration requires multiple steps, and hence errors accumulate and the convergence may change. Thus, we set $i = n$. Knowing that $(1 + O(d))^n = O(1)^6$ and inserting Equation (5.13) into Equation (5.12), we obtain:

$$T_\lambda = \prod_{j=0}^{i-1} \exp(-\tau_j d) + O(d)O(1) \tag{5.14}$$

$$= \prod_{j=0}^{i-1} (1 - \tau_j d + O(d^2)) + O(d). \tag{5.15}$$

Our last step is to show that the first term on the right side of Equation (5.15) also converges linearly with respect to d. In the course of this section, we omit the presence of the term $O(d)$ in Equation (5.15) for clarity. Let us define the set K as the set of indices j for which $1 - \tau_j d = 0$. The size of K is denoted as $|K| = k$. We also define $\overline{K}$ as the set of indices j for which $1 - \tau_j d \neq 0$, and $|\overline{K}| = i - k$. Equation (5.15) can be written as:

$$T_\lambda = \left(\prod_{j \in \overline{K}} (1 - \tau_j d + O(d^2))\right)\left(\prod_{j \in K} O(d^2)\right) \tag{5.16}$$

$$= \left(\prod_{j \in \overline{K}} (1 - \tau_j d + O(d^2))\right) O(d^{2k}). \tag{5.17}$$

Because $1 - \tau_j d \neq 0$ for $j \in \overline{K}$:

$$T_\lambda = \left(\prod_{j \in \overline{K}} (1 - \tau_j d)\left(1 + \frac{O(d^2)}{1 - \tau_j d}\right)\right) O(d^{2k}). \tag{5.18}$$

---

[6] Let $(1 + ax)^{b/x}, a, b \in \mathbb{R}^+$ and $x \to 0$:

$$(1 + ax)^{\frac{b}{x}} = \exp\left(\frac{b}{x} \log(1 + ax)\right)$$
$$= \exp\left(\frac{b}{x}(ax + O(x^2))\right)$$
$$= \exp(ab + O(x))$$
$$= 1 + ab + O(x) = O(1).$$

Recall that $d = D/n$:

$$(1 + O(d))^n = (1 + O(d))^{D/d} = O(1).$$

From the definition of big O notation, $1/(1 - \tau_j d) = O(1)$; hence:

$$T_\lambda = \left(\prod_{j \in \overline{K}} (1 - \tau_j d) \left(1 + O(1)O(d^2) \right) \right) O(d^{2k}) \tag{5.19}$$

$$= \left(\prod_{j \in \overline{K}} (1 - \tau_j d)(1 + O(d^2)) \right) O(d^{2k}) \tag{5.20}$$

$$= \left(\prod_{j \in \overline{K}} 1 - \tau_j d \right) (1 + O(d^2))^{i-k} O(d^{2k}). \tag{5.21}$$

Note that $k \neq 0$ implies that at least one of the terms $1 - \tau_j d = 0$. Thus, the code accumulating the value of T, $T = T * (1 - t_j * d)$, will return $T = 0$. Because we want to recover the approximation errors for the general case ($T_\lambda \neq 0$), we set $k = 0$ in Equation (5.21), and $i = n$ for the same reasons as previously stated:

$$T_\lambda = \left(\prod_{j=0}^{n-1} 1 - \tau_j d \right) (1 + O(d^2))^n. \tag{5.22}$$

Using the fact that $(1 + O(d^2))^n = 1 + O(d)$ and $(1 + O(d))^n = O(1)$:

$$T_\lambda = \left(\prod_{j=0}^{n-1} 1 - \tau_j d \right) (1 + O(d)) \tag{5.23}$$

$$= \left(\prod_{j=0}^{n-1} 1 - \tau_j d \right) + O(d) \left(\prod_{j=0}^{n-1} (1 + O(d)) \right) \tag{5.24}$$

$$= \left(\prod_{j=0}^{n-1} 1 - \tau_j d \right) + O(d)(1 + O(d))^n \tag{5.25}$$

$$= \left(\prod_{j=0}^{n-1} 1 - \tau_j d \right) + O(d)O(1) \tag{5.26}$$

$$= \left(\prod_{j=0}^{n-1} 1 - \tau_j d \right) + O(d). \tag{5.27}$$

We finish our derivation[7] by adding the previously omitted $O(d)$:

$$T_\lambda = \left(\prod_{j=0}^{n-1} 1 - \tau_j d\right) + O(d) + O(d) \tag{5.28}$$

$$= \left(\prod_{j=0}^{n-1} 1 - \tau_j d\right) + O(d). \tag{5.29}$$

Equation (5.29) concludes the asymptotic convergence of the inner integral. Knowing that T_λ converges linearly, our next step is to approximate the outer integral.

Approximation of the Outer Integral

Let $\tilde{T}_i$ be the approximation of $T(\lambda_i)$. We write $T(\lambda_i) = T_i = \tilde{T}_i + O(d)$ and $C_i = C(id)$. In typical volume rendering implementations, the outer integral is also approximated using a Riemann summation. Thus:

$$I(x, y) = \sum_{i=0}^{n-1} C(id)\tau(id)T_i d + O(d) \tag{5.30}$$

$$= \sum_{i=0}^{n-1} C_i \tau_i d \left(\tilde{T}_i + O(d)\right) + O(d) \tag{5.31}$$

$$= \sum_{i=0}^{n-1} C_i \tau_i \tilde{T}_i d + \sum_{i=0}^{n-1} C_i \tau_i d O(d) + O(d). \tag{5.32}$$

---

[7] Let $(1 + ax^2)^{b/x}, a, b \in \mathbb{R}^+$ and $x \to 0$:

$$(1 + ax^2)^{\frac{b}{x}} = \exp\left(\frac{b}{x} \log(1 + ax^2)\right)$$
$$= \exp\left(\frac{b}{x}(ax^2 + O(x^4))\right)$$
$$= \exp\left(abx + O(x^3)\right)$$
$$= 1 + abx + O(x^3)$$
$$= 1 + O(x).$$

Recall that $d = D/n$:

$$(1 + O(d^2))^n = (1 + O(d^2))^{\frac{D}{d}}$$
$$= 1 + O(d).$$

Because both τ_i and C_i are bounded, one can write $C_i \tau_i d O(d) = O(d^2)$ and $\sum_i O(d^2) = nO(d^2) = \frac{D}{d} O(d^2) = O(d)$. The above equation can be rewritten as:

$$I(x, y) = \sum_{i=0}^{n-1} C(id)\tau(id)d\tilde{T}_i + O(d). \tag{5.33}$$

$$= \sum_{i=0}^{n-1} C_i \tau_i d \left(\prod_{j=0}^{n-1} 1 - \tau_j d \right) + O(d). \tag{5.34}$$

We have now showed that the dominant error when considering step size in the VRI is of order $O(d)$. In other words, when decreasing the step size by half, the error should be reduced by half as well.

5.4   VERIFICATION ALGORITHMS

The derivation shown in the previous section gives us two important pieces of information: (1) it shows that the the volume rendering discretization does indeed converge to the correct solution; and (2) it shows how fast it converges to that solution. Based on this information, we explain two verification procedures: *convergence analysis* and *order of accuracy*.

5.4.1   CONVERGENCE ANALYSIS

The first verification algorithm tests whether the errors are reducing toward zero, regardless of the speed of convergence. Specifically, we will measure the errors between consecutive images generated using different numbers of samples; then we will plot these errors to evaluate whether the curve generated does indeed converge to zero. Suppose we want to verify the convergence of a sequence of images I_i with $d_{i+1} = \frac{1}{2} d_i$. As we have seen in the previous section, the approximation for the solution F at resolution i and $i + 1$ can be written, respectively, as:

$$F(x, y) = I_i(x, y) + O(d_i^k)$$
$$= I_i(x, y) + \beta d_i^k + \text{HOT}, \tag{5.35}$$
$$F(x, y) = I_{i+1}(x, y) + O(d_{i+1}^k)$$
$$= I_{i+1}(x, y) + \beta d_{i+1}^k + \text{HOT}, \tag{5.36}$$

where (x, y) are the screen coordinates of the pixel being evaluated and HOT are High Order Terms, which we will assume to be negligible. Now, we subtract Equation (5.36) from Equation (5.35) to eliminate the unknown "true image" F:

$$0 = (I_{i+1}(x, y) + \beta d_{i+1}^k) - (I_i(x, y) + \beta d_i^k) \tag{5.37}$$
$$0 = I_{i+1}(x, y) - I_i(x, y) + \beta d_{i+1}^k - \beta d_i^k. \tag{5.38}$$

Thus, the convergence order k can be computed by evaluating the errors involved in the subtraction of consecutive images:

$$
\begin{aligned}
e_i(x, y) = I_{i+1}(x, y) - I_i(x, y) \quad &= \quad -\beta d_{i+1}^k + \beta d_i^k \qquad\qquad (5.39) \\
&= \quad \beta(1 - (1/2)^k)d_i^k. \qquad (5.40)
\end{aligned}
$$

We use the L_∞ norm to compute the maximum error among all pixels:

$$
\begin{aligned}
E_i \quad &= \quad \sup_{x,y}(e_i(x, y)) \\
&= \quad \sup_{x,y}(|I_{i+1}(x, y) - I_i(x, y)|) = \beta(1 - (1/2)^k)d_i^k. \qquad (5.41)
\end{aligned}
$$

Thus, the observed convergence is again computed by taking logarithms from both sides. We then write $y = \log \beta(1 - c^k)$ to hide the dependency of the term in k and determine y and k via least-squares:

$$
\begin{aligned}
\log E_i \quad &= \quad \log \beta(1 - (1/2)^k)d_i^k \qquad\qquad (5.42) \\
&= \quad \log \beta(1 - (1/2)^k) + k \log d_i \qquad (5.43) \\
&= \quad y + k \log d_i. \qquad\qquad\qquad (5.44)
\end{aligned}
$$

Equation (5.44) shows us how to compute the convergence rate using only the images obtained from the VRI approximation and consequently avoiding any bias and/or limitations introduced by simple manufactured solutions or numerical approximations using reference images. We have generated sequences of images based on the refinements in the following section. The steps are shown in Algorithm 4. Note that one could use a high-precision numerical approximation as a reference image, for example, and measure the convergence toward that image. The main disadvantage of this approach is that it might mask errors that appear in the reference image itself.

5.4.2   ORDER OF ACCURACY

The main difference between the order of accuracy method and convergence analysis is that the order of accuracy measures the convergences toward a *manufactured solution*. As the name suggests, a manufactured solution is an analytical solution to the VRI that can be used as the ground truth. The advantage of using a manufactured solution over convergence analysis is that the former measures whether we are converging at the right speed to the right solution. Depending on the type of error introduced, convergence analysis may converge at the right speed to a solution that is not the correct one.[8] If follows that we will need a solution to the VRI. Although the VRI does not have a solution in the general case, we can devise solutions that are simple enough so

---

[8]As a simple exercise, imagine a scenario where a bug is inadvertently introduced into the I/O procedure: while reading the input scalar field G, the software accidentally changes it to G'. A convergence analysis test will miss this problem. The issue is that convergence analysis uses G' to generate images, and not to build a ground truth. Thus, the method can change the input data without having any real impact on the convergence of approximation errors.

Algorithm 4 An algorithm for volume rendering verification via convergence analysis.

Verify(G, τ, C, d_0, M)

    ▷ Let G be an $N \times N \times N$ scalar field
    ▷ Let τ be a transfer function defining the extinction coefficient
    ▷ Let C be a transfer function defining the color
    ▷ Let d_0 be the initial step size
    ▷ Let M be the number of images to be generated for testing convergence
    ▷ Let VolumeRendering($\cdots$) be the under verification

1    $F_0 \leftarrow$ VolumeRendering(G, τ, d_0)
2   **for** $i \leftarrow 1$ **to** M
3   **do** $d_i = \frac{1}{2}d_{i-1}$
4           $F_i \leftarrow$ VolumeRendering(G, τ, C, d_i)
5           $E_i = \max_{x,y} |F_{i-1}(x, y) - F_i(x, y)|$
6   Linear regression of $\{E_i\}_{i=1}^M$ using Equation (5.44):

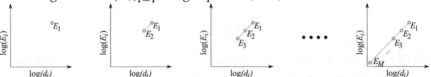

The practitioner evaluates whether the behavior is close enough to the expected.
The slope of the line is k, the expected convergence rate.

---

we can find a solution but complex enough to reveal potential problems with the technique under verification.

    Let us assume that a solution $F(x, y)$ for the VRI is known. Then, the procedure described next is equivalent to the method of manufactured solutions [2]. The solution F can be written as:

$$F(x, y) = I(x, y) + O(d^k) = I(x, y) + \beta d^k + \text{HOT}, \tag{5.45}$$

where I is the approximated image, d is the sample distance, and $\beta \in \mathbb{R}$ is a constant, multiplicative factor that is not a function of the dataset. An important assumption is that the HOT, or "higher-order terms," are small enough that they do not affect the convergence of order $k \in \mathbb{R}$, i.e., high-order derivatives of F must have a negligible impact on the asymptotic convergence of I [47]. This formulation implies that not all solutions F are suitable for verification purposes, only those for which the HOT are negligible. The errors can be written as:

$$e(x, y) = I(x, y) - F(x, y) \approx \beta r^k. \tag{5.46}$$

One can evaluate the convergence for all pixels in the image using L_2, L_∞, or other norms. Henceforth, we adopt the L_∞ norm because it provides a rigorous and yet intuitive way of eval-

---

Algorithm 5 An algorithm for volume rendering verification via order of accuracy.

VERIFY(G, τ, C, d_0, M)

    ▷ Let G be an $N \times N \times N$ scalar field

    ▷ Let τ be a transfer function defining the extinction coefficient

    ▷ Let C be a transfer function defining the color

    ▷ Let I be the analytical solution of the VRI for $\{\tau, C, G\}$

    ▷ Let d_0 be the initial step size

    ▷ Let M be the number of images to be generated for testing convergence

    ▷ Let VOLUMERENDERING($\cdots$) be the under verification

1 **for** $i \leftarrow 1$ **to** M

2   **do** $d_i = \frac{1}{2}d_{i-1}$

3        $F_i \leftarrow$ VOLUMERENDERING(G, τ, C, d_i)

4        $E_i = \max_{x,y} |I(x, y) - F_i(x, y)|$

5 Linear regression of $\{E_i\}_{i=1}^M$ using Equation (5.44):

The user evaluates whether the behavior is close enough to the expected.
The slope of the line is k, the expected convergence rate.

---

uating errors. It tells us that the maximum image error should decay at the same rate k. Mathematically, the error is then:

$$E = \sup_{x,y}(e(x, y)) = \sup_{x,y}(|I(x, y) - F(x, y)|) = \beta d^k. \tag{5.47}$$

We denote individual images (and the respective errors) by a subscript i. For each image I_i, we first calculate the supremum of the absolute difference $\sup_{x,y} (|F(x, y) - I_i(x, y)|)$. We then compute the observed convergence rate k by taking logarithms of both definitions of E and solving the resulting equations for $\log(\beta)$ and k in a least-squares sense:

$$\begin{aligned} \log E_i &= \log \sup_{x,y} (|F(x, y) - I_i(x, y)|) \\ &= \log(\beta) + k \log(d_i). \end{aligned} \tag{5.48}$$

The system of equations has as many equations as the number of images and calculated errors. Algorithm 5 shows the steps for volume rendering verification via order of accuracy.

5.5 APPLICATION EXAMPLES

Let us illustrate the concepts shown in previous sections with practical examples. Suppose we are implementing the volume rendering algorithm and we want to verify it. What are the steps involved in the process of verification? The first is to decide what is the expected behavior of the errors introduced by the technique under verification. If our implementation uses the standard discretization of the VRI using the Riemann sum, then Section 5.3 tells us that we should expect the errors involved in our implementation to decay linearly. Next, we need to find analytical solutions that will be used as the ground truth during the verification procedure. We use the following analytical solution:

$$I(x, y) = 1 - \exp\left(\frac{\cos(\cos(xy))}{\cos(xy)} - \frac{1}{\cos(xy)} \right).$$

See Table 5.1 for an overview of all the input data used to generate this solution. Note that this solution is applicable for the combination of transfer functions, scalar field, domain, and ray parameterization shown in Table 5.1. Thus, the input for our volume rendering technique must be identical to that used for producing the analytical solution. Last, we will use Algorithm 5 to

Table 5.1: The analytical solution I of the VRI for the set of parameters shown on the left

Input	Value
domain of interest	$[0, 1]^3$
ray length	$D = 1$
ray parameter	λ
ray position	$\mathbf{x}(\lambda) = (x, y, \lambda)$
scalar field	$s(x, y, z) = z \cos(xy)$
extinction coefficient tf	$\tau(s) = \sin(s)$
color tf	$C(s) = 1$
analytical solution I	$I(x, y) = 1 - \exp\left(\frac{\cos(\cos(xy))}{\cos(xy)} - \frac{1}{\cos(xy)} \right)$

generate a plot containing the error decay. If the slope of the line is equivalent to the expected order of accuracy (in our scenario $k = 1$), then we consider the technique to be verified.

To show the possible unexpected behaviors one can expect while using this verification procedure, we applied our algorithm to three volume rendering techniques freely available for download and testing: Voreen [36], VTK Fixed-Point Volume Ray Caster, and VTK Volume Ray Caster. All algorithms presented use the standard discretization of the VRI, which means that we should expect linear convergence.

5.6 RESULTS

In what follows, we describe the results of applying the convergence and order of accuracy tests to three mature implementations of solutions of the VRI. Here we will test how errors behave

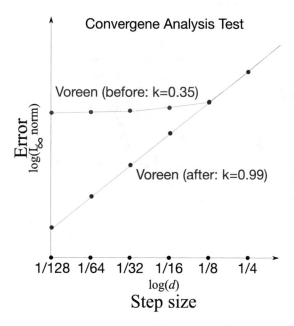

Figure 5.6: Voreen convergence analysis test. The red curve shows that the early ray termination acceleration technique does have an effect on the numerical accuracy of the method. The blue curve shows the results of our test without early ray termination.

(converge) by progressively increasing the number of samples points per ray, i.e., reducing the ray distance. There are other ways to evaluate the error incurred by discretization, such as evaluating the errors as the dataset is refined. In this case, another theoretical analysis is required to reveal how the expected errors decay as the dataset is refined.

5.6.1   VOREEN

Voreen (Volume Rendering Engine) uses the power of GPU (Graphics Processing Units) to implement real-time volume rendering algorithms. Voreen is a robust, trustworthy, and widely used volume rendering package. We applied convergence analysis test (see Section 5.4.1) to version 2.6.1 of Voreen. Figure 5.6 shows the result of applying the convergence analysis test. The horizontal axis shows the size of the step size d. We progressively reduce the step size by half in order to compute the errors generated. The red and blue curves show the convergence before and after, respectively, fixing problems that prevented the correct convergence. The red curve converges nicely during the first two steps of refinement, but then the maximum error remains constant, regardless of the ray sampling. The constant error is interesting because high sampling rates are used to obtain high precision images. Consider this: if we were to generate an image to be used as a gold standard, we would have used high sampling rates in order to produce low-error images.

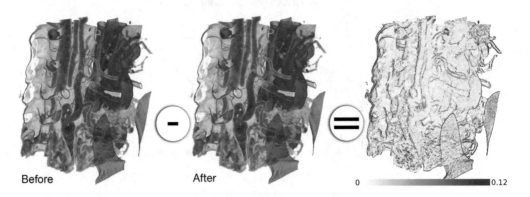

Figure 5.7: Vertebra dataset. Although the "before" and "after" images look alike, they are slightly off, which results in an error spread throughout the whole image. Source: Etiene et al. [11].

The red curve shows us that the approximation errors will not diminish beyond a certain point as we refine our ray. In the best case scenario, we will waste computational resources.

The nonlinear convergence was due to the early ray termination (ERT). The ERT is a technique commonly used in volume rendering implementations as a tradeoff between speed and accuracy. As the ray traverses the volume, the technique measures the "opacity levels" of the domain. If the domain becomes "too opaque," then the system halts the ray traversal, saving time. The issue is "too opaque:" a hardcoded threshold that developers believe to be small enough to avoid incurring errors, but large enough to save computer power. In the end, convergence analysis measures precisely how much error one should expect and the maximum step size that makes sense for a particular threshold. By simply reducing the value of the ERT threshold, we were able to obtain the expected convergence for ray refinement (shown in the blue curve). The slope of the curve is nearly the expected value $k = 1$, which is good enough for purposes of verification.

Voreen had additional problems that were not caught by convergence analysis of the ray step refinement d, but they were revealed by convergence analysis of progressive refinement of the input scalar field G (see Etiene et al. [11] for details). In a nutshell, the uncovered problem resulted in a stretching of the whole image due to boundary interpolation issues. Figure 5.7 illustrates the problem. The image on the left shows the incorrect result, whereas the middle image shows the expected result, and the right image shows the difference between the correct and incorrect. Notice that the error is small, but everywhere in the image. In this case, the problem does not affect the visual quality of the image (as both the correct and incorrect image are indistinguishable). Nevertheless, this is not a guarantee that these kinds of bugs will not have an effect on interpretation and/or other quantitative results.

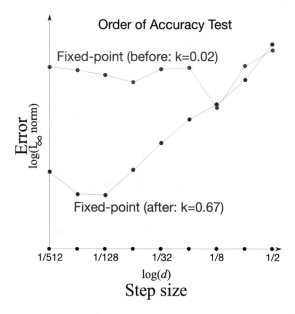

Figure 5.8: Fixed-point Volume Ray Caster order of accuracy test. The red curve shows our initial results. Because the ray traverses only half of the input volume, the rendered images do not convergence to the expected results. The blue curve shows the results after we fixed the ray traversal procedure.

5.6.2 VTK FIXED-POINT VOLUME RAY CAST

The VTK Fixed-Point Volume Ray Cast technique implements the standard Riemann discretization of the VRI, but it does so using fixed-precision arithmetic. It uses 15-bits precision, instead of the standard 32 and 64 floating-point precision, to generate images. The advantage of the fixed-precision approach is that it can lead to better performance. All computation becomes integer operations. So, we should expect that images generated using fixed point will be more sensitive to small changes in the input values. The standard floating-point types can handle small step sizes (i.e., large sampling rate), but a 15-bit fixed precision will be able to handle only relatively larger step sizes. Figure 5.8 shows the results obtained. The red curve shows the convergence rate obtained after our first test. We obtained nearly constant error decay ($k = 0.02$) when we expected a linear decay. Although the fixed-precision may explain the issue, we expect it to play a role only after the step size d becomes too small to be represented in fixed-precision. In other words, the approximation errors should decrease for large values of d, and increase when d is too small (errors due to 15-bit fixed-point precision). After investigating the reason for this deviation, we found that depending on the pixel, camera, and ray position, some rays might cross only half of the volume instead of the full volume, which can be seen in Figure 5.9. The darker artifacts appear because the ray has not traversed the whole volume at those locations. After we fixed those

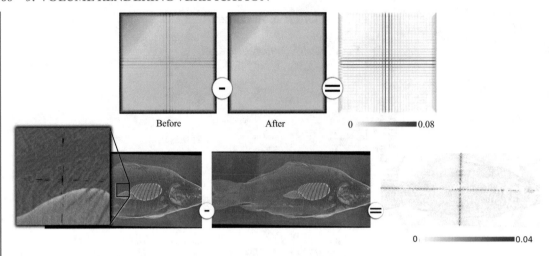

Figure 5.9: Result of using the Fixed-Point Volume Ray Caster volume rendered with a synthetic dataset. The "+" pattern is evident. Darker pixels belong to regions where the ray traverses only half of the volume, preventing convergence. The middle images show the result of our modified version of the VTK. The "+" pattern can also be seen in real-world datasets. Source: Etiene et al. [11]

problems, we obtained the convergence curve shown in blue. The error convergence is now much closer to the expected linear convergence.[9]

5.6.3   VTK VOLUME RAY CASTER

The VTK Volume Ray Caster module also implements the standard Riemann discretization of the VRI using floating-point arithmetic. One (correctly) expects better precision, and hence lower errors, when compared to fixed-point arithmetic. The Volume Ray Caster algorithm produced the best results of the three tested algorithms. It produced nearly linearly converging sequences when refining the step size. Figure 5.10 shows that the red curve—the results of the volume rendering technique before introducing any modification—has almost the expected convergence. The only red flag raised was related to the dataset convergence test (outside the scope of this book). The test revealed a small inconsistency in the total number of steps necessary to traverse the input volume. By correcting the inconsistency, one obtains the blue curve.

5.7   DISCUSSION

The VTK Volume Ray Caster was the technique that performed better in convergence analysis tests. The VTK Fixed-Point Volume Ray Caster was the only technique that produced visible ar-

---

[9]The convergence curve is almost linear if f we discard the last three samples, which are potentially the result of the 15-bit fixed-precision arithmetic.

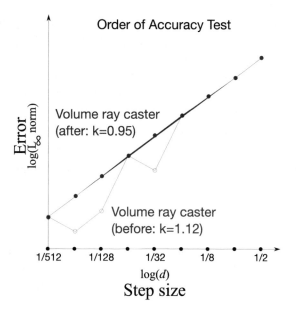

Figure 5.10: Volume Ray Caster order of accuracy test. No issues were revealed with the proposed test.

tifacts. Nevertheless, the three techniques under verification manifested—either through testing the convergence of the sample distance or through another convergence test—problems that prevented us from obtaining the expected theoretical convergence rate. Notice that the fact that the convergence rate was not obtained does not imply that the algorithm or implementation contains problems. Rather, it serves as a warning that an issue is present in some part of the verification pipeline. A concrete example is the early ray termination optimization (ERT). The ERT is not a bug, but a deliberate design decision to accelerate the performance of the volume rendering implementation. In this sense, our verification test measures the amount of numerical errors introduced by this optimization, which can then be compared against the performance gains. In other words, in this case, the verification procedure at least provided tools to measure the tradeoff between quality and speed. It is also interesting to notice that all the techniques that were tested are very stable and well tested. Verification through convergence analysis is sensitive enough to detect small deviations from the ideal results.

We have applied verification to understand how real, production-quality implementations behaved. We were looking for *unknown* issues within the tested code. A slightly different approach, known in software engineering as *mutation testing*, is to introduce changes to the code and then see whether the verification procedure is able to detect them. In other words, we will use the verification technique to track *known*, injected issues. Of course, this kind of test is not meant to be exhaustive, but to shed light on some of the potential limitations of the verification procedure. The results of this test are illustrated in Table 5.2. Out of the 12 issues, 8 were detected, i.e., the value of slope of the convergence curve substantially deviated from $k = 1$. These eight include incorrect opacity accumulation and incorrect ray increment. Four issues escaped detection, namely, issues #9 to #12. By investigating each of the issues, we were able to identify the reasons the verification test failed. For instance, in the case of issue #10, a problem was inserted in the green and blue component of the pixel color, not in the red component. As it turns out, the verification procedure was tailored to use only the red channel, instead of all three, which prevented it from detecting the bug. By increasing coverage of the verification procedure, the problem can be detected. Issue #11 turned out to be a dead-code; therefore, changing it caused no changes in the final result.

5.8   CONCLUSIONS

The verification procedure shown in this chapter can be a powerful tool to developers interested in implementing volume rendering techniques. The errors introduced by the discretization of the VRI lie at the core of the verification procedure. By looking at how these errors behave as the input parameters of the implementation under verification change, one can gain valuable insights into the code correctness. To motivate the reader to put the concepts shown in this chapter into practice, we designed two challenges. In the first challenge, users will try to implement a correct emission-absorption model. Go to `http://tiagoetiene.github.com/verifier` to get more information and take the challenge. As users type in their code, the implementation is automatically verified in real-time, and the results are plotted, so users know how well they are doing. If the user's result is close to the expected, the verification is declared successful. In the second challenge, users are responsible for implementing the verification routine. We provide a verified volume rendering implementation. The responsibility of the user is to implement the verification routines and provide all the input data necessary to verify the code.

Table 5.2: Sensitivity of the verification procedure. On the right, we show the volume rendering algorithm and the location in which a bug was introduced (marked as #n). On the bottom, the first row shows the images rendered *with* the introduced bugs. The second row shows the error between the exact (leftmost image in the first row) and the generated images. The 12 issues shown below were injected in the VTK Volume Ray Cast technique. Source: Etiene et al. [11].

#	Issue	Detected
1	Incorrect opacity accumulation	Yes
2	Incorrect ray increment	Yes
3	Small changes to ERT	Yes
4	Piecewise constant τ	Yes
5	Incorrect matrix–point multiplication	Yes
6	Incorrect evaluation of interpolant	Yes
7	Uninitialized pixel center offset	Yes
8	Incorrect coefficients computation 1	Yes
9	Incorrect coefficients computation 2	No
10	Incorrect color lookup	No
11	Incorrect matrix multiplication	No
12	Incorrect loop range	No

VOLUME RENDERING
for each pixel
  do Find pixel center (#7)
    Transform rays to voxels space (#5, #11)
    for each step along the ray (#12)
      do Compute interpolant coeff (#8, #9)
      Interpolate scalar values (#6)
      Compute color and opacity (#4, #10)
      Compositing (#1)
      Increment sample position (#2)
      Check for ERT (#3)

4×10^{3} 0.3

Exact solution 1 2 3 4 5 6 7 8 9 10 11 12

CHAPTER 6

Conclusion

The term *verification* has become ubiquitous in both the computer science and engineering communities as referring to a process that somehow convinces the user that verified tools, whether those be circuits, algorithms, implementations, etc. are more safe, accurate, or complete than other tools that have not been verified. Although the term verification has a common root usage within both communities, it has evolved to mean something specific to each subarea of computer science and engineering. For instance, within computer science, the verification of a circuit denotes either the exhaustive testing or proof that under all possible inputs, the circuit will produce the correct (specified) outputs. Similarly, for software, verification refers to how well an implementation represents the behavior of its specification under all possible inputs. Within the engineering world, verification takes on a different, more nuanced meaning. One assumes that there exists an "exact solution" or representation resulting from the solution of a mathematical system of equations. In all but the most trivial circumstances, this exact solution is not attainable, and approximate solutions must be formed. The process of quantifying how well a numerical scheme or representation approximates the exact solution is referred to as verification. Verification may involve looking at how well (or quickly) an approximate solution converges (in an appropriate norm) to the exact solution, or identifying features or invariants of the solution that should be maintained regardless of the approximate representation. As visualization models, algorithms, and implementations lie at the interface between the CS and CS&E communities, what does it mean to produce *verifiable* visualizations, and equally importantly, does it matter if visualizations are verifiable? These are distinct questions from, but intimately related to, questions of perception and visual representation efficacy. The purpose of this book was to articulate clearly what the mathematical verification of visualization models, algorithms, and implementations means in the contexts provided above; to articulate how and why verification matters to both the "producers" of visualizations (i.e., the visualization community); and the "consumers" (i.e., the sciences and engineering), and to set forth examples of verifying some commonly used visualization schemes.

Bibliography

[1] F. J. Ascombe. Graphs in Statistical Analysis. *The American Statistician*, 27(1):17–21, February 1973. DOI: 10.1080/00031305.1973.10478966. 5

[2] I. Babuska and J. Oden. Verification and validation in computational engineering and science: basic concepts. *Computer Methods in Applied Mechanics and Engineering*, 193(36-38):4057–4066, 2004. DOI: 10.1016/j.cma.2004.03.002. 2, 3, 19, 20, 60

[3] D. Betounes. *Partial Differential Equations for Computational Science: With Maple and Vector Analysis*. Springer, 1998. 3

[4] E. V. Chernyaev. Marching Cubes 33: Construction of topologically correct isosurfaces. Technical Report CN/95-17, Institute for High Energy Physics, 1995. 15

[5] L. Custodio, T. Etiene, S. Pesco, and C. Silva. Practical considerations on Marching Cubes 33 topological correctness. *Computers & Graphics*, 37(7):840 – 850, 2013. DOI: 10.1016/j.cag.2013.04.004. 15

[6] J. W. Demmel. *Applied Numerical Linear Algebra*. Society for Industrial and Applied Mathematics, Philadelphia, PA, 1997. 4

[7] C. Dietrich, C. Scheidegger, J. Schreiner, J. Comba, L. Nedel, and C. Silva. Edge transformations for improving mesh quality of marching cubes. *Visualization and Computer Graphics, IEEE Transactions on*, 15(1):150–159, January 2009. DOI: 10.1109/TVCG.2008.60. 42

[8] M. J. Dürst. Re: Additional reference to "Marching Cubes". *SIGGRAPH Computer Graphics*, 22(5):243, October 1988. DOI: 10.1145/378267.378271. 14

[9] H. Edelsbrunner, J. Harer, V. Natarajan, and V. Pascucci. Morse-smale complexes for piecewise linear 3-manifolds. In *SCG '03: Proceedings of the nineteenth annual symposium on Computational geometry*, pages 361–370, New York, 2003. ACM. DOI: 10.1145/777792.777846. 3

[10] H. Edelsbrunner and E. P. Mücke. Simulation of simplicity: a technique to cope with degenerate cases in geometric algorithms. *ACM Transactions on Graphics*, 9(1):66–104, 1990. DOI: 10.1145/77635.77639. 4

[11] T. Etiene, D. Jonsson, T. Ropinski, C. Scheidegger, J. L. Comba, L. G. Nonato, R. M. Kirby, A. Ynnerman, and C. T. Silva. Verifying volume rendering using discretization error analysis. *IEEE Transactions on Visualization and Computer Graphics*, 20(1):140–154, 2014. DOI: 10.1109/TVCG.2013.90. 51, 54, 64, 66, 69

[12] T. Etiene, L. Nonato, C. Scheidegger, J. Tienry, T. Peters, V. Pascucci, R. Kirby, and C. Silva. Topology verification for isosurface extraction. *IEEE Transactions on Visualization and Computer Graphics*, 18(6):952–965, June 2012. DOI: 10.1109/TVCG.2011.109. 45

[13] T. Etiene, C. Scheidegger, L. Nonato, R. Kirby, and C. Silva. Verifiable visualization for iso-surface extraction. *Visualization and Computer Graphics, IEEE Transactions on*, 15(6):1227–1234, Nov 2009. DOI: 10.1109/TVCG.2009.194. 36, 41

[14] M. Friendly. A brief history of data visualization. In *Handbook of Data Visualization*, Springer Handbooks of Computational Statistics, pages 15–56. Springer Berlin Heidelberg, 2008. DOI: 10.1007/978-3-540-33037-0_2. 7, 8

[15] A. Globus and E. Raible. Fourteen ways to say nothing with scientific visualization. *Computer*, 27(7):86–88, July 1994. DOI: 10.1109/2.299418. 17

[16] A. Globus and S. Uselton. Evaluation of visualization software. *SIGGRAPH Computer Graphics*, 29(2):41–44, May 1995. DOI: 10.1145/204362.204372. 4, 16

[17] P. Godefroid. Software model checking: The verisoft approach. *Formal Methods in System Design*, 26(2):77–101, 2005. DOI: 10.1007/s10703-005-1489-x. 4

[18] P. Godefroid. Random testing for security: blackbox vs. whitebox fuzzing. In *RT '07: Proceedings of the 2nd international workshop on Random testing*, pages 1–1, New York, 2007. ACM. DOI: 10.1145/1292414.1292416. 4

[19] J. Heer. Socializing visualization. In *In CHI 2006 Workshop on Social Visualization*. Citeseer, 2006. 8

[20] L. Hesselink, F. Post, and J. van Wijk. Research issues in vector and tensor field visualization. *Computer Graphics and Applications, IEEE*, 14(2):76 –79, March 1994. DOI: 10.1109/38.267477. 4

[21] J. Humar. *Dynamics of Structures*. A.A. Balkema Publishers, 2002. 26

[22] IBM. OpenDX. http://www.research.ibm.com/dx. 3

[23] C. Johnson, R. Moorhead, T. Munzner, H. Pfister, P. Rheingans, and T. S. Yoo. *NIH/NSF Visualization Research Challenges Report*. IEEE, 2006. DOI: 10.1109/MCG.2006.44. 3

[24] R. Kirby and C. Silva. The need for verifiable visualization. *IEEE Computer Graphics and Applications*, 28(5):78–83, September 2008. DOI: 10.1109/MCG.2008.103. 2, 4, 11

[25] R. M. Kirby and Z. Yosibash. Solution of von-karman dynamic non-linear plate equations using a pseudo-spectral method. *Computer Methods in Applied Mechanics and Engineering*, 193/6-8:575–599, 2004. DOI: 10.1016/j.cma.2003.10.013. 24

[26] G. Klein, K. Elphinstone, G. Heiser, J. Andronick, D. Cock, P. Derrin, D. Elkaduwe, K. Engelhardt, R. Kolanski, M. Norrish, T. Sewell, H. Tuch, and S. Winwood. sel4: formal verification of an os kernel. In *SOSP '09: Proceedings of the ACM SIGOPS 22nd symposium on Operating systems principles*, pages 207–220, New York, 2009. ACM. DOI: 10.1145/1629575.1629596. 4

[27] T. Lewiner. *http://www.matmidia.mat.puc-rio.br/tomlew*, 2012 (accessed July 24, 2012). 15

[28] T. Lewiner, H. Lopes, A. W. Vieira, and G. Tavares. Efficient implementation of Marching Cubes' cases with topological guarantees. *Journal of Graphics Tools*, 8(2):1–15, 2003. DOI: 10.1080/10867651.2003.10487582. 15

[29] A. Lopes and K. Brodlie. Improving the robustness and accuracy of the marching cubes algorithm for isosurfacing. *Visualization and Computer Graphics, IEEE Transactions on*, 9(1):16–29, January 2003. DOI: 10.1109/TVCG.2003.1175094. 15

[30] W. E. Lorensen and H. E. Cline. Marching Cubes: A high resolution 3D surface construction algorithm. *SIGGRAPH Computer Graphics*, 21(4):163–169, August 1987. DOI: 10.1145/37402.37422. 11, 14

[31] C. Lündstrom, P. Ljung, A. Persson, and A. Ynnerman. Uncertainty visualization in medical volume rendering using probabilistic animation. *Visualization and Computer Graphics, IEEE Transactions on*, 13(6):1648–1655, Nov 2007. DOI: 10.1109/TVCG.2007.70518. 50

[32] S. R. Marschner and R. J. Lobb. An evaluation of reconstruction filters for volume rendering. In *Proceedings of the Conference on Visualization '94*, VIS '94, pages 100–107, Los Alamitos, CA, 1994. IEEE Computer Society Press. DOI: 10.1109/VISUAL.1994.346331. 17

[33] N. Max. Optical models for direct volume rendering. *Visualization and Computer Graphics, IEEE Transactions on*, 1(2):99–108, June 1995. DOI: 10.1109/2945.468400. 47, 48

[34] B. H. McCormick. Visualization in scientific computing. *SIGBIO Newsletter*, 10(1):15–21, March 1988. DOI: 10.1145/43965.43966. 9

[35] M. Meissner. The Volume Library @ONLINE. `http://volvis.org`, June 2014. 17

[36] J. Meyer-Spradow, T. Ropinski, J. Mensmann, and K. H. Hinrichs. Voreen: A rapid-prototyping environment for ray-casting-based volume visualizations. *IEEE Computer Graphics and Applications.*, 29(6):6–13, November/December 2009. DOI: 10.1109/MCG.2009.130. 62

[37] C. Montani, R. Scateni, and R. Scopigno. A modified look-up table for implicit disambiguation of Marching Cubes. *The Visual Computer*, 10(6):353–355, 1994. DOI: 10.1007/BF01900830. 15, 16

[38] B. Natarajan. On generating topologically consistent isosurfaces from uniform samples. *The Visual Computer*, 11(1):52–62, 1994. DOI: 10.1007/BF01900699. 15

[39] T. S. Newman and H. Yi. A survey of the marching cubes algorithm. *Computers & Graphics*, 30(5):854–879, 2006. DOI: 10.1016/j.cag.2006.07.021. 13, 15

[40] G. Nielson. On marching cubes. *Visualization and Computer Graphics, IEEE Transactions on*, 9(3):283–297, July 2003. DOI: 10.1109/TVCG.2003.1207437. 15

[41] G. M. Nielson and B. Hamann. The asymptotic decider: resolving the ambiguity in Marching Cubes. In *Proceedings of the 2nd Conference on Visualization '91*, VIS '91, pages 83–91, Los Alamitos, CA, 1991. IEEE Computer Society Press. DOI: 10.1109/VISUAL.1991.175782. 15, 31

[42] W. L. Oberkampf and C. J. Roy. *Verification and Validation in Scientific Computing*. Cambridge University Press, 2010. DOI: 10.1017/CBO9780511760396. 19

[43] Paraview. http://www.paraview.org. 3

[44] S. G. Parker and C. R. Johnson. SCIRun: a scientific programming environment for computational steering. page 52, 1995. 3

[45] M. Pauly and M. Gross. Spectral processing of point-sampled geometry. In *SIGGRAPH '01: Proceedings of the 28th annual conference on Computer graphics and interactive techniques*, pages 379–386, New York, 2001. ACM. DOI: 10.1145/383259.383301. 3

[46] P. J. Roache. Building pde codes to be verifiable and validatable. *Computing in Science and Engineering, Special Issue on Verification and Validation*, pages 30–38, September/October 2004. DOI: 10.1109/MCSE.2004.33. 19

[47] C. J. Roy. Review of code and solution verification procedures for computational simulation. *Journal of Computational Physics*, 205(1):131–156, 2005. DOI: 10.1016/j.jcp.2004.10.036. 60

[48] W. Schroeder, K. Martin, and B. Lorensen. *The Visualization Toolkit An Object-Oriented Approach To 3D Graphics*. Kitware, 2003. 3

[49] W. Schroeder, K. M. Martin, and W. E. Lorensen. *The Visualization Toolkit (2nd ed.): An Object-oriented Approach to 3D Graphics*. Prentice-Hall, Inc., Upper Saddle River, NJ, 1998. 42

[50] J. R. Shewchuk. Delaunay refinement algorithms for triangular mesh generation. *Computational Geometry*, 22(1-3):21 – 74, 2002. DOI: 10.1016/S0925-7721(01)00047-5. 4

[51] S. T. Tony Hey and K. T. (Editors). *The Fourth Paradigm: Data-intensive scientific discovery*. Microsoft Research, 2007. DOI: 10.1007/978-3-642-33299-9_1. 1

[52] E. R. Tufte. *The Visual Display of Quantitative Information*. Graphics Press, Cheshire, CT, 1986. 5, 8

[53] E. R. Tufte. *Visual Explanations: Images and Quantities, Evidence and Narrative*. Graphics Press, Cheshire, CT, 1997. DOI: 10.1063/1.168637. 8

[54] S. Uselton, G. Dorn, C. Farhat, M. Vannier, K. Esbensen, and A. Globus. Validation, verification and evaluation. In *VIS '94: Proceedings of the conference on Visualization '94*, pages 414–418, Los Alamitos, CA, 1994. IEEE Computer Society Press. DOI: 10.1109/VISUAL.1994.346285. 4

[55] The VisTrails Project. http://www.vistrails.org. 3

[56] Volume ray casting. Licensed under creative commons attribution-share Alike 3.0 via Wikimedia commons. http://commons.wikimedia.org/wiki/File:Volume_ray_casting.png, August 2014. 49

[57] M. Wattenberg. Baby names, visualization, and social data analysis. In *IEEE Symposium on Information Visualization, 2005*, pages 1–7, October 2005. DOI: 10.1109/INFVIS.2005.1532122. 8

[58] R. Wenger. *Isosurfaces: Geometry, Topology, and Algorithms*. An A K Peters Book. Taylor & Francis, 2013. 30

[59] R. B. Wilhelmson, B. F. Jewett, C. Shaw, L. J. Wicker, M. Arrott, C. B. Bushell, M. Bajuk, J. Thingvold, and J. B. Yost. A study of the evolution of a numerically modeled severe storm. *International Journal of Supercomputing Applications and High Performance Engineering*, 4(2):20–36, July 1990. DOI: 10.1177/109434209000400203. 7, 8

[60] P. L. Williams and S. P. Uselton. Metrics and generation specifications for comparing volume-rendered images. *The Journal of Visualization and Computer Animation*, 10(3), July 1999. DOI: 10.1002/(SICI)1099-1778(199907/09)10:3%3C159::AID-VIS205%3E3.0.CO;2-A. 4

Authors' Biographies

TIAGO ETIENE

Tiago Etiene received his computer science and M.S. degrees from the University of São Paulo in 2006 and 2008, respectively, and his Ph.D. degree from the University of Utah in 2013. His research on the field of verifiable visualizations serves as the basis for this book. In addition, his research interests include geometry processing, computer graphics, scientific and information visualization, and related areas. He is a research scientist at Modelo Inc. working with cutting-edge geometry processing tools tailored to geosciences applications.

ROBERT M. KIRBY

Robert M. (Mike) Kirby received an M.S. degree in applied mathematics, an M.S. degree in computer science, and a Ph.D. degree in applied mathematics from Brown University, Providence, RI, in 1999, 2001, and 2002, respectively. He is currently a Professor of Computing and Associate Director of the School of Computing, University of Utah, Salt Lake City, where he is also an Adjunct Professor in the Department of Bioengineering and Mathematics and a member of the Scientific Computing and Imaging Institute. His current research interests include scientific computing and visualization.

CLÁUDIO T. SILVA

Cláudio T. Silva is a professor of computer science and engineering and data science at New York University. Claudio's research lies in the intersection of visualization, graphics, data analysis, and geometric computing, and recently he has been interested in urban data and sports analytics. He has published over 220 journal and conference papers, and he is an inventor of 12 U.S. patents. He received a B.S. in mathematics from the Federal University of Ceará (Brazil) in 1990, and a Ph.D. in computer science from State University of New York at Stony Brook in 1996. He has held positions in academia and industry, including at AT&T, IBM, Lawrence Livermore, Sandia, and the University of Utah. He is an active member of the research community, has participated in more than 100 international program committees, served on the editorial boards of multiple journals, including the *IEEE Transactions on Visualization and Computer Graphics, ACM Transactions on Spatial Algorithms and Systems,* and *IEEE Transactions on Big Data,* and served as general and papers co-chair of a number of conferences, including the IEEE Visualization. Silva's work has received a number of major awards. In 2013, he was elected an Institute of Electrical and Electronics Engineers fellow and in 2014 he won the IEEE Visualization Technical Achievement Award. He helped develop MLB.com's Statcast player tracking system, which won the Alpha Award for Best Analytics Innovation/Technology at the 2015 MIT Sloan Sports Analytics Conference. In 2015, he was elected Chair of the IEEE Technical Committee on Visualization and Computer Graphics.